T0229822

1 URBAN ANALYSIS

A STUDY OF CITY STRUCTURE WITH
SPECIAL REFERENCE TO SUNDERLAND

URBAN ANALYSIS

A STUDY OF CITY STRUCTURE
WITH SPECIAL REFERENCE
TO SUNDERLAND

by B. T. ROBSON

Lecturer in Geography
in the University of Cambridge

CAMBRIDGE
AT THE UNIVERSITY PRESS 1971

CAMBRIDGE UNIVERSITY PRESS
Cambridge, New York, Melbourne, Madrid, Cape Town, Singapore,
São Paulo, Delhi, Dubai, Tokyo, Mexico City

Cambridge University Press
The Edinburgh Building, Cambridge CB2 8RU, UK

Published in the United States of America by Cambridge University Press, New York

www.cambridge.org
Information on this title: www.cambridge.org/9780521099899

First published 1969
Reprinted 1971
Re-issued 2010

A catalogue record for this publication is available from the British Library

Library of Congress Catalogue Card Number: 68-25086

ISBN 978-0-521-07272-4 Hardback
ISBN 978-0-521-09989-9 Paperback

To Glenna

CONTENTS

Contents

EDITORS' FOREWORD

The Syndics of the Cambridge University Press have decided to institute a series of books and monographs to be entitled *Cambridge Geographical Studies*; and have appointed us the editorial board.

Our intention is to include in the series books which either describe and illustrate the new ideas and techniques now reshaping geographical work, or books which otherwise embody the results of important new research. The works published will not be textbooks; but it is hoped that the series as it develops will cover all aspects of geography.

B. H. FARMER
A. T. GROVE
Cambridge E. A. WRIGLEY
31 January 1968

PREFACE

The cultural revolution in geography may not have been of Chinese proportions, but it has wrought profound effects on the nature of the subject, in terms both of the quantification of its approach and of the widening range of its subject matter. Increasingly, human geographers are including aspects of sociological material within their field of interest. This book attempts to show the relevance of sociology to a geographical approach to urban areas and, for such an urban social geography, the revolutionary little red book is still Park, Burgess and McKenzie's *The City*. While its ideas can be extensively criticized, they are none the less constantly stimulating and still provide perhaps as much insight into the nature of the urban area as any more recently propounded concepts. The ecologists' interest in spatial associations and areal patterns is very close to the geographical approach, and this book explores the relevance of the old ecological models and tests the validity of new ones. That the old models have to be altered and new approaches experimented with is a reflection of recent changes in the social structure of towns. In trying to work towards new models and new concepts, the geographer must increasingly incorporate some of the data and ideas of sociology into his own peculiar spatial methodology.

What in essence is attempted here is a geographical method of analysing the social structure of a single town and the use of the results of this analysis in a spatial examination of one facet of the town's sociology, namely the development of attitudes to education amongst parents of boys who were about to sit the 11-plus examination. The principal sources of material which are drawn upon are enumeration district data from the 1961 Census, data from Valuation Lists at three dates, and the results of a questionnaire survey in seven selected areas of the town.

The substantive data relate to the town of Sunderland and the choice of this town calls for some comment. To some extent, the reasons for its selection were personal since North-East England is an area well known to the author, but there were also strong practical reasons underlying its selection. First, the size of the town—just under 200,000 population in 1961—provided a unit large enough to ensure well-developed internal functional segregation without being beyond the resources of a single researcher. Second, its administrative boundaries correspond closely to the built-up area, so that official data could be used more confidently than if, as is often the case, large areas of housing had lain outside the boundaries. Third, and of great importance, was the fact of the town's social composition. With its large proportion of working class population and its extensive development of local authority housing, it was possible to study a wide variety of types of working class areas and to examine the effects of council housing on the nature of the town's structure. The preponderance of workers in the traditional industries of shipbuilding and heavy manufacturing might make the town less representative of the so-called 'new' working class, but was ideal for the research design of this book.

In collecting the raw data, I was met with great friendliness and ready assistance by many officials of the Sunderland Local Authority, especially the staffs of the Planning Office, the Education Office and the Public Library and Museum. I should particularly like to thank Mr R. Oliver and the staff of the Rating Office for the help and hospitality which they gave me in making the Valuation Lists of the town available.

Discussion with fellow members of the Inter-University Census Tract Committee helped to clarify many of the problems of urban analysis, and in particular the suggestions of Miss Elizabeth Gittus, of the University of Newcastle upon Tyne, were of great help. For help in programming the component analysis I am indebted to Miss Sylvia Lutkins of the Statistics Department, Aberystwyth, and for helpful comment on certain statistical problems, to Mr Jeffrey Round, of the Department of Economics, Aberystwyth. The editors of this series made some useful comments on the first draft of this book for which I am grateful.

In the writing of this book, I am aware of the debt which I owe to Professors E. G. Bowen and C. Kidson and my ex-colleagues at the Department of Geography, University College of Wales, Aberystwyth, who provided a pleasing and stimulating atmosphere in which to work. The cartography of the diagrams speaks for the skill of the technical staff of that same Department and for their help I am most grateful.

For financial help in buying enumeration district data for Sunderland from the General Register Office, I should like to acknowledge St Catharine's College, Cambridge and the Trustees of the Lake Fund of the Department of Geography, Cambridge University.

I am grateful for permission to reproduce the following diagrams: Figs. 3.11, 3.12, 3.14 to 3.19 and 3.21 to 3.23 which first appeared in an article in *Urban Studies*, vol. III (1966); and Figs. 4.7 to 4.10 which first appeared in a chapter in E. G. Bowen (ed.), *Geography at Aberystwyth*, University of Wales Press, 1968.

PART ONE

METHODOLOGY

CHAPTER I

HUMAN ECOLOGY AND THE GEOGRAPHY OF TOWNS

Attitudes to towns seem never to be neutral. Some may be violently antago-
nistic, as in the long intellectual history of revulsion against all things urban
and the Rousseauesque idealization of the rural. Some may be enthusiastic-
ally favourable, as in the praises sung to the civilizing effects of urbanization.
The city has called forth its reformers, its prophets, its proselytizers, its
planners, its poets and writers. But whatever the attitude, there is no doubting
the fascination that towns offer to the interested observer or to the scholar.
The flux of changing skylines and altering land uses; the intimate intermingling and the patterning of urban landscapes; the seemingly purposive
movement of goods and people; all offer scope for reflection or for study.

Urban studies have grown, in recent years, to encompass a wide variety of
disciplines which each contribute their individual viewpoints to a common
interest in aspects of the urban area. A reflection of this process of convergence has been the growth of a number of interdisciplinary research and
teaching centres devoted to the study of the city. In America, the Harvard-
M.I.T. Joint Center for Urban Studies and the Chicago Center for Urban
Studies are just two from no fewer than nine such bodies. Their British
equivalents are the London Centre for Urban Studies established in 1958 and
the Birmingham Centre for Urban Studies which was started in 1966. Like
the rise of Isard's 'regional science', a complete field of urban studies has
emerged, delineated in terms of a given type of area rather than the con-
ventional systematic disciplinary distinction which is based on a defined
range of subject matter. The world of these urban study centres links the
expertise of economists, sociologists, geographers, planners, traffic engineers
and a host of other experts who ask variants of the same questions: what is the
city, how does it function, what are its sustaining mechanisms, what are its
implications for human behaviour and economic development? Although a
long way from resolving such questions, different scholars have tried to
provide answers within the frameworks of legalistic, historical, economic,
sociological and psychological concepts.

There are perhaps two major reasons for this synthesizing tendency within the urban area. First, the city is so much an extension of man himself, so much his own creation, that he is unable to stand apart from it to specify and pigeonhole its component aspects. The city is more than a field of study. In highly compressed form, it is man's home, a palimpsest of local history, a standing monument to man's creativeness and the nexus of a system of economic interdependencies. His personal involvement makes the complex intermeshing of the city's multifarious strands all the more evident. This may be a matter of degree rather than of kind, since the interconnections of any set of phenomena with a range of other sets can only be ignored at peril, but the dense texture of urban phenomena intensifies the linkage of the many mutually dependent elements and makes it conceptually more difficult to abstract any one from the total functioning entity.

The second and more important reason is the application of new ranges of techniques and new types of approach to the analysis of the urban area. The convergence upon common urban problems, which has characterized aspects of a number of disciplines, has been hastened by the development of methods of handling areal data with more powerful statistical techniques. Perhaps for the first time, a common language for communication between the different specialists has been provided by these techniques. As the discussion of the following chapter will show, the application of methods of manipulating areal data has highlighted the parallel problems and the areas of convergence between those disciplines which contribute to urban studies. Techniques never provide a substitute for a philosophical orientation, but they do have profound implications for the way in which scholars look at the data and at the problems which they handle. Just as Darwin's concepts in the nineteenth century or quantum physics in the twentieth had ramifications throughout a range of disciplines in both the physical and social sciences, so the development of statistical theory and the application of probabilistic laws, which have derived from it, have profoundly altered the methodology of many disciplines today. Within urban studies, the contributing disciplines have been drawn together and the fusion has produced some healthy cross-fertilization. In the same way, within the whole subject of geography, which contributes one of the strands of urban studies, the development of theoretical-deductive approaches in place of empirical-inductive approaches has brought together the once separate fields of economic, urban and transport geography into the realm of locational analysis.[1]

Such welding of disciplines and parts of disciplines has made ever more intimate the interrelationships between them, and nowhere is this truer than in the research nodes on the borders of the more traditional fields. In this

[1] A view which is admirably elaborated in *The science of geography*, Report of the Committee on Geography, National Academy of Sciences—National Research Council (Washington, D.C., 1965).

chapter sociology and geography, two of the disciplines which contribute to urban studies, will be examined to show how a blending of their respective viewpoints can provide a valid framework for the analysis of the social composition of the urban area.

THE NATURE OF GEOGRAPHY

Geography has long been rife with internal discussion of its philosophy and content. The many strands which have contributed to the history of geographical thought have been admirably surveyed by Hartshorne in his compendious *Nature of Geography*.[1] More recently he has restated his perspective on the subject in terms of a traditional approach, seeing geography as 'that discipline that seeks to describe and interpret the variable character from place to place of the earth as the world of man'.[2] His emphasis on 'areal differentiation' reflects a continuing theme within geography: a theme which has placed regional identification and description at the core of the subject. A variation of this approach is the elucidation of the relationships between man and his environment. This too has a long pedigree within the world of geography whether its proponents have styled themselves as determinists, possibilists or probabilists.

However, in spite of Hartshorne's plea that definitions of the subject have to be couched in terms of what has gone before and that delineation of its methodology can only be made by reference to what the founding fathers said and did, modern geography has increasingly developed its definitions without reference to the controversies and discussions of the past half century or more. The availability of more powerful descriptive and analytical techniques—especially of correlations of spatial distributions and the various techniques for analysing spatial pattern—has helped to develop a definition of the subject in terms of its viewpoint as a science of distribution. By defining geography as the study of spatial relationships one effectively encompasses a great deal of past geography and provides a rationale for defining and attacking research problems. As many writers have pointed out, such a cryptic definition in no way provides an exclusive domain for geography since distributional aspects play a greater or lesser part in a large number of systematic subjects such as economics, botany or medicine. But such overlap is as healthy as it is expected. There is no necessary virtue in trying to delineate a completely 'watertight' discipline—even if this were possible. As it is, concentration on locational interrelationships leads the geographer to the functional interconnections which are expressed through them and so to a

[1] R. Hartshorne, *The nature of geography* (Chicago, 1939).
[2] *Idem, Perspective on the nature of geography* (Chicago, 1959), p. 47.

5

consideration of the processes and concepts of whatever systematic field his interests most closely impinge upon. The geographer's ultimate concern is therefore marginally, if at all, different from the related systematic scientist's. The urban geographer, for example, may study the distribution of crime and his analysis must inevitably lead to a consideration of the same processes and factors as those which interest the criminologist. The important distinction is that by starting with a locational or spatial reference point, the geographer is likely to unearth hypotheses and processes which might not have been revealed by an analysis starting within the systematic subject itself.

An example which comes to mind is the geographical approach to disease, which is developing a considerable body of literature. Howe, for example, looks at the distribution of deaths from various causes in Great Britain[1] and, while he makes no attempt to interpret the patterns which emerge, the distributions do give rise to interesting speculation and provide a base for detailed research into the factors underlying some of the spatial patterns. The high incidence of stomach cancer in parts of Wales, for example, suggests the working of some environmental factors which might be included in an aetiology of the disease. Howe also notes the classic case of John Snow in nineteenth-century London. On the basis of maps of the detailed distribution of deaths in the East End, Snow argued in vain that cholera was spread by infected water supplies. Surreptitious removal of the suspect water pump confirmed his ideas and the confirmation was a validation of the relevance of a distributional viewpoint as a source of fruitful hypotheses.

This distributional definition of the subject leaves unanswered a great many detailed questions as to concepts and methods of approach in geography, but, without going into detail at this point, an idea of basic theoretical orientation will emerge in the following discussion of the scope of social geography in urban areas.

Evidently, if one defines geography in terms of its methods of approach, the subject matter to which this approach is applied will be determined by the particular systematic branch with which one is concerned. A full definition of the subject, in other words, depends on delineating what one studies as well as how one studies it. In the case of social geography, one would deal with the spatial arrangements and the functional interrelationships of human groups. To this extent there is a great deal of overlap between urban social geography and urban sociology. In each case, the aggregate problem is similar: the analysis and understanding of urban social structure. We can define social structure in broad terms to mean a system of patterned relationships between groups of persons. The geographer will approach such a study from the point of view of the distribution both of the groups themselves and

[1] G. M. Howe, *National atlas of disease mortality in the United Kingdom* (London, 1963).

6

also those artefacts of the groups which might reflect the functional patterns of group relations. In all cases, the existence of recurring or systematic pattern will be taken as a possible consequence of the relations which exist between the human groups. Spatial pattern, in other words, is taken to reflect functional processes. In studying the distribution of types of houses or of the incidence of social disorders, for example, the aggregate problem is illuminated only in so far as such distributions can reveal something of the functional aspects of the economic and social characteristics of the groups of people with which one deals.

A concern with macro-sociology of this sort is a view with which many sociologists would disagree. Despite Durkheim's insistence on the need to study society not the individual, and the need to explain social facts with reference to other preceding social facts,[1] a great deal of sociological concern now centres on the individual, who is seen as being more tangible and more 'real' than 'society'. A few years ago, for example, there was an interesting controversy as to the explanation of race relations within Britain. Richmond drew upon Adlerian psychology and such concepts as Adorno's 'authoritarian personality' to develop an approach which stressed the psychological explanation of prejudice.[2] He considered that sociological facts were merely a stimulus which may control the expression of latent psychological prejudice. Banton, by contrast, approached race relations from a more sociological standpoint by stressing the importance of the social situation in which race contacts occurred.[3] While giving primacy to situational variables, he regarded personality variables as purely circumstantial factors.

Richmond's approach might be said to constitute an example of 'psychological sociology' whereas Banton's would be 'social psychology'. Each approach is concerned both with group and individual characteristics, either as dependent or independent variables. However, the closest sociological parallel with the geographical approach which has been outlined is to be found in the work of the human ecologists. As Schnore argues, the human ecologists can be regarded as 'macro-sociologists' in that both their dependent and independent variables are *group* characteristics.[4] The history of ecological thought is a fascinating one and not so familiar to geographers as it should be since the similarity between aspects of the two subjects is great. Except for a casual acquaintance with the 'concentric circle' model and the 'sector' model much of the theoretical structure of human ecology is ignored

[1] E. Durkheim, *The rules of sociological method*, English translation (Glencoe, Illinois 1950), p. 110.
[2] A. H. Richmond, *Colour prejudice in Britain* (London, 1954).
[3] M. P. Banton, *White and Coloured : the behaviour of British people towards coloured immigrants* (London, 1959).
[4] L. F. Schnore, 'The myth of human ecology', *Sociological Inquiry*, XXI (1961), 128–39, reprinted in L. F. Schnore, *The urban scene : human ecology and demography* (New York, 1965), ch. 2.

by geographers. Before considering the parallels between the two subjects therefore, it would be as well to consider briefly something of the changing methodology of human ecology throughout its relatively short span of development.

THE DEVELOPMENT OF ECOLOGICAL THOUGHT

'Ecology' is a much over-used and frequently misused term.[1] It is essential at the outset to make a cursory distinction between its various applications. As originally used, the term was applied to the physical sciences of botany and zoology in their study of the relationships of plants and animals with their physical environment. By the early years of this century, plant ecology in particular had become a well-developed field with a large body of research material to its credit and a clearly elaborated theoretical background. The application of certain of the principles and processes of biological ecology to the field of social science gave rise to human or social ecology; a field which was first developed in America in the early part of this century, associated particularly with work done at Chicago. The development of human ecology in fact derived from a number of sources.[2] Alihan notes, for example, the dependence of the early ecologists on the theories of the human physiologist Child, on the empirical writings of the land economist Hurd, but most particularly on the writings of such plant ecologists as Clements.[3] It was the use of biological analogies in the study of human populations which provided the initial impetus to the theory of human ecology.

If plant ecology, animal ecology and human ecology provide three fields in which the term has been applied, a fourth field is in general ecology. Here the term is stretched to its fullest extension to include the vast and complex pattern of the whole of nature. General ecology is thus holistic in that, as Taylor has declared, it is the study of '*all* relations of *all* organisms to *all* their environment'.[4] While, in a rather different guise, there has been a revival of interest among geographers in such a general ecology, it is the much more limited concept of human ecology which is of interest here. Where the term 'human ecology' or simply 'ecology' is used, it refers to the ideas and concepts which have been developed from the original formulations

[1] For an excellent discussion of the breadth of the various interpretations see O. D. Duncan, 'Human ecology and population studies', in P. M. Hauser and O. D. Duncan (eds.), *The study of population : an inventory and appraisal* (Chicago, 1959), pp. 678–81.

[2] Interestingly, these include the writings of the geographers la Blache, Ratzel and Brunhes.

[3] M. A. Alihan, *Social ecology : a critical analysis* (New York, 1938), ch. 5. The specific borrowings are from the following: C. M. Child, *The physiological foundations of behaviour* (New York, 1924); R. M. Hurd, *Principles of city land values* (New York, 1911); F. E. Clements, *Plant succession : an analysis of the development of vegetation* (Washington, 1916).

[4] W. P. Taylor, 'What is ecology and what good is it?', *Ecology*, XVII (1936), 335.

8

of the Chicago school. While the Chicago concepts are not by their nature restricted to the urban area, but could be applied to rural areas outside the city proper, it happens that the great bulk of the writings of the Chicago ecologists has been concerned almost exclusively with urban phenomena. This is partly an accident of history in that the founders of human ecology worked within the context of Chicago, a city which embodied *par excellence* the rapid expansion of urban America at the turn of the century. But a more important explanation than this is the fact that human ecology has been concerned with dynamic processes which were most evident and best studied within the context of the changes and stresses which rapid urban expansion gave rise to.

In the early part of this century Chicago was the centre of a remarkable, almost frantic, outburst of interest in the city, guided by the triad of Park, Burgess and McKenzie. Park was originally a journalist whose initial interest in the city was purely pragmatic; he commented, 'I expect that I have actually covered more ground tramping about in cities in different parts of the world, than any other living man. Out of all this I gained, among other things, a conception of the city, the community, and the region, not as a geographical phenomenon merely, but as a kind of social organism.'[1] He went to Chicago University in 1914, and, influenced by the writings of Darwin and the plant ecologists, began to develop a theory of the city based on an analogy between plant communities and human communities. The early years of the development of human ecology were devoted to factual surveys of Chicago, which was used as the human laboratory for the ecologists. Detailed mapping of a host of economic and social phenomena by Park's students formed the basis of his theoretical exegesis. His ideas were elaborated in a series of papers which laid the ground work and provided the springboard for his many followers. Beginning with an article written in 1916,[2] he later elaborated and developed his doctrine in *The Introduction to the Science of Society* which was written in collaboration with Burgess in 1921. The first formal presentation of the collected views of the Chicago school was in *The City*, which was published in 1925, by which time the human ecologists had produced an impressive array of ecological monographs.

The starting point of Park's ecological theory was Darwin's concept of the web of life: the intimate interrelationship between organism and organism

[1] Quoted in R. E. Park, *Human Communities*, the collected papers of Robert Ezra Park (Glencoe, Illinois, 1952), II, 5.
[2] R. E. Park, 'The city: suggestions for the investigation of human behaviour in the urban environment', *Am. J. of Sociol.* xx (1916), 577–612. This, and other of Park's writings, have been reprinted in three volumes of his collected works issued by the Free Press of Glencoe under the titles, *Race and Culture* (1950); *Human Communities* (1952) and *Society* (1955). The best account of the foundation of the Chicago school is by Burgess, one of its principal figures, in E. W. Burgess and D. J. Bogue (eds.), *Contributions to urban sociology* (Chicago, 1964), pp. 2–14.

and between organism and environment. Since man is an organic creature, Park argued that he is subject to the general laws of the organic world. This laid the basis for his use of biological analogy. However, as Park recognized, the analogy is imperfect since man is subject to impulses other than the basic need for survival. Therefore to accommodate both the sub-social and the social factors, he built his conceptual framework of human ecology around a fundamental distinction between the two concepts of 'community' and 'society' which were to be considered as distinct aspects of human life; so distinct in fact that the study of community should fall within the purview of ecology while that of society should be included in the subject matter of social psychology. This distinction was based upon two different levels of human activity; the biotic and the cultural. The biotic gave rise to the community and was based on the sub-social forces of competition. At this level, people were to be regarded as 'individuals' lacking distinctively social attributes and therefore subject to the same impulses and forces as were plants or animals in their struggle for existence and for the acquisition of the most favourable circumstances in which to live. The cultural level, on the other hand, gave rise to society and was based on the strictly social processes of communication and consensus in which people become 'persons' with social attributes. It was the operation of these cultural processes which distinguished man from other organic elements in nature. Society, characteristic of the increasing sophistication of man, was therefore seen as a superstructure lying above the more basic competitive biotic level of community. The realm of ecological concern was therefore to be found in the community which was produced by inevitable 'natural' forces similar to those applying to plants and animals.

At the biotic level, the various processes recognized by the plant ecologists could therefore be translated into human terms. First, and most fundamental, was the concept of competition. Man competed for limited space and for access to the most desirable location for his residence and for his business activities. Such competitive activity was reflected in land values which, through the price mechanism, sorted out like types of person into similar sorts of areas. It was this process which accounted for the segregation of the Central Business District, the areas of commerce, and the residential areas of similar sorts of people. Different types of people were segregated in terms of their ability to pay the various rental levels. The slum area represented an area of minimum choice and so collected a population which was homogeneous in terms of its economic competency, even though in ethnic terms it may be very heterogeneous. Competition thus led to segregation of like types of persons and like types of business and commercial activity.

A second process was that of dominance, which is one of the fundamental concepts of the plant ecologists. Within different types of plant associations, one species exerts a dominant influence in that it controls the environmental conditions which encourage or discourage other types of species. In the beech climax, for example, the beech tree is the dominant element since its height and foliage determine the amount of light which is filtered through to lower layers and so determine the types of plants which are found at these lower layers. Typically, one finds an association of beech with such plants as bluebells, which flower at a season when the beech trees have lost their foliage and thus permit the maximum amount of light penetration. In the same way within the city, the Central Business District forms the dominant element within the whole complex of the urban area since competition between business concerns to locate in the area of maximum accessibility gives rise to the pattern of rising land values closer to the city centre which in turn affects the disposition of other elements within the urban complex. Again, within local areas, particular types of activity exert dominance. The high status area is dominated by higher income people who resist the encroachment of lower income peoples; the industrial areas are dominated by those industries which, by their noxious character, repel residential development. It is in this sense that the slum is seen as an area of minimum choice since it attracts the residue of those elements of the city which, with their low level of economic competency, are effectively unable to compete for more advantageous sites.

Dominance, in the ecological parlance, was intimately connected with the concepts of invasion and succession. Again, the analogy with the plant world is close. Plants, by their activity in changing the micro-environment in which they live, create conditions in which other, less tolerant plants are able to thrive, and these other species begin to invade the environment, eventually to establish themselves as dominant elements and, as the process continues, form part of a succession of dominant elements which, in plant ecology, tends towards a climax association. This process of invasion and succession was applied to human communities in relation to the invasion of residential areas by commercial and business undertakings and of higher status residential areas by lower income groups. In Chicago, with its highly varied ethnic composition, invasion of one type of ethnic area by people of a different ethnic stock was also common and provided admirable examples of the process of invasion and the establishment of dominance by the invading group.

These, very briefly, were some of the applications of ecological theory to the human community. The value and the stimulus of the Park concepts was that they provided a holistic view of the mechanisms underlying the functioning of the urban area. Everyday observation and earlier factual studies had long suggested that similar types of human and economic phenomena tended to form more or less segregated areas within cities. Booth's studies of London, for example, were a superb example of nineteenth-century work on urban areas which gave descriptive evidence of this tendency. In Chicago, the work of Jane Addams and others at Hull House provided useful insights into the slum conditions of parts of the city. But until Park's theoretical suggestions such knowledge was fragmentary and incoherent. As George Ponderevo, in H. G. Wells' *Tono-Bungay*, said of London, it was possible to imagine 'a kind of theory' of the city since he could discern 'lines of an ordered structure out of which it has grown' and 'detect a process that is something more than confusion of casual accidents'. Before Park, knowledge of urban structure was as groping as Ponderevo's, but Park gave a rationale and a theoretical underpinning to subsume such observations into an ordered and, the ecologists maintained, a predictable set of categories.

There were other important concepts which were derived from the ecological analogy. The first, related to the process of dominance, was the concept of the gradient. It was in developing this concept that the ecologists drew upon Child's physiological writings. Child stressed the 'axiate' pattern of complex organisms in which, for example, the brain develops in the most active region and other organs 'arise in definite order along the axes'. With evolutionary growth he postulated an increasing differentiation of an organism's component parts and an increasing concentration of control at the point of dominance.[1] The ecologists noted that analogously within the city there was a gradient of land values which declined outwards from a peak at the functional centre of the city, and that many other social phenomena assumed this gradient form influenced by the sifting and sorting effects of land values. Income groups were shown to be graded outwards from the centre. So too were such rates of social disorder as crime and mental illness. Second, related to the process of segregation, was the concept of the natural area. Natural areas were viewed by some ecologists as being delineated by patterns of land utilization, which were moulded by transportation, industrial organization, topography and the like, and which were 'the unplanned, natural product of the city's growth'.[2] On the other hand, other ecologists (such as McKenzie) saw them as cultural rather than physical features which were composed of homogeneous areas of race, language, income and occupation. Whatever the defining characteristics, the natural areas were seen as small areas of homogeneity produced through the operation of competition and the segregating effects of land values. The high degree of internal homogeneity within them was seen to make them act as sub-systems within the larger community.

[1] Child, *Physiological foundations of behaviour*.
[2] H. W. Zorbaugh, 'The natural areas of the city', in E. W. Burgess (ed.), *The urban community* (Chicago, 1925), p. 222.

These underlying concepts were brought together in an overall spatial model by Burgess' well-known theory of city growth. In studying Columbus, Ohio, in 1923, McKenzie had suggested the division of the city into concentric circles which expressed the form of the city's growth. Burgess applied this idea to Chicago with his concentric-zone model which, with its five component rings, was the logical spatial expression of the ecological principles of central dominance, segregation, invasion and succession.[1] Development outwards was accompanied by the differentiation of the successive rings of new urban growth and the invasion by different elements of the older inner rings of the city. The characteristics of the five concentric rings illustrate in more detail the spatial expression of the ecological processes. The first ring, the Loop in Chicago, Burgess considered to be the Central Business District. The second ring was the zone in transition in which invasion or incipient invasion by commerce and business led to a fall in residential desirability and the existence of a cheek-by-jowl mixture of land uses: industry, commerce and business intermingled with high-density, highly subdivided residential accommodation, occupied by the poor and the undesirable. At one time having been on the periphery of the city many of the houses in this zone were substantial, but their value was only a potential value in that the area was about to become attractive for business. Repairs and renovations were thus unprofitable and the decay which was allowed to set in created conditions in which maximum profit could be realized in terms of subdivisions of housing and letting accommodation as rooms. It was therefore in this area that the undesirable elements of society were found with their social practices, such as prostitution or gambling, which were socially necessary but officially denigrated. It was, too, the area of invasion of new incoming ethnic groups who found in it cheap accommodation and a lack of strict enforcement of social controls. The third zone, the 'zone of working-men's homes', contained the small inexpensive frame houses of the working class. Burgess considered this a zone of second-generation settlement, thus illustrating the process of successive invasion. Immigrants who had assimilated American social values more completely and had become more affluent in the process of assimilation, typically moved outwards from the less desirable and cheaper accommodation of the zone in transition into this third zone. The move itself was a process by which their assimilation was confirmed and consolidated. The fourth zone, the 'zone of better residences', continued this process of rising status with decreasing centrality. It was an area of essentially middle-class population, of white collar employees and professional people, and composed largely of single-family residences of

[1] E. W. Burgess, 'The growth of the city: an introduction to a research project', *Publ. Am. Sociol. Soc.* XVIII (1924), 85–97. Reprinted in G. A. Theodorson (ed.), *Studies in human ecology* (Evanston, 1961), pp. 37–44.

exclusive character. The final fifth zone, the 'zone of commuters', was a suburban area, with the suburbs contained either as part of the city or as separate towns. Functional distance determined the distance from the Central Business District that commuters lived, and Burgess considered that a journey of some 30–60 minutes defined the outer limits of this predominantly affluent zone.

Such were the processes and patterns upon which the ecology of the Chicago school was based. The effects of Park and Burgess' teaching were far-reaching. From the early period of fact-finding about the distribution of social phenomena within the city, students went on to complete an array of impressive surveys of urban life and structure. They included studies of whole communities such as McKenzie's *The neighbourhood: a study of Columbus, Ohio* (1923), studies of certain types of area within Chicago such as Louis Wirth's *The Ghetto* (1928), and Zorbaugh's *The gold coast and the slum* (1929), and studies of types of social groups or individuals, relating their social aspects to their environmental setting, such as Anderson's *The Hobo* (1923), Trasher's *The Gang* (1927), Clifford Shaw's *Delinquency Areas* (1929), or Cressey's *The Taxi Dance-Hall* (1932). The great majority of such studies, of which these are only a few of the most important, began as doctoral dissertations within the university.[1]

Empirical criticism

While the 1920s and early 1930s were thus a time of infectious enthusiasm and productive effort in the realm of human ecology, the middle and late 1930s saw an increasing amount of criticism of the ecological stance. This criticism was directed on both the empirical and the theoretical fronts. At an empirical level, the writings of Edith Abbott suggested that, applied to Chicago, Burgess' model of urban growth and many of the concepts of the ecologists were 'theories that seem to be purely theoretical and not realistic'.[2] Burgess and all the ecologists recognized that the model was purely conceptual and that certain disturbing influences, such as topography and the disposition of lines of communication, did distort the theoretically concentric circles when applied to specific towns. Within Chicago, the line of the lake front, for example, attracted high-class residences which reached into the central parts of the Loop. But Abbott questioned the fruitfulness of the very use of the zonal model in the light of such distortions of the patterns.

On the basis of a wider selection of cities, Davie in 1938 took up the empi-

---

[1] Many of the works which came from this productive period, in addition to more recent ecological writings, have been collected in two readers: Burgess and Bogue (eds.), *Contributions to urban Sociology*, and Theodorson (ed.), *Studies in human ecology*.
[2] E. Abbott, *The tenements of Chicago, 1908–1935* (Chicago, 1936).

rical criticism of the period.[1] Looking at some twenty cities within the United States, and in particular at the town of New Haven, he suggested the overwhelming importance of lines of communication and of the disposition of industry within towns as factors which tended to shape the residential patterning of the urban area. In particular he suggested that industry tended to be located near the lines of water or rail transport 'wherever in the city this may be—and it may be anywhere', and that low-grade housing was found near these industrial and transport areas. From his 'general principles governing the distribution of utilities' he concluded that there was 'no universal pattern, not even of an "ideal" type'.[2] Hard on the heels of Davie's criticism were the suggestions of Homer Hoyt who, as a land economist, worked outside the sphere of human ecology itself, but whose ideas had a close bearing on the general tendency for empirical doubts to be thrown on the ecologists' models at this period. In 1939 Hoyt produced the Federal Housing Administration's report, *The structure and growth of residential neighborhoods in American cities*. On the basis of rental data culled from a large number of cities, Hoyt evolved a new model of urban structure which differed from the zonal pattern of Burgess. Tracing the movement of the high status residential neighbourhood, he emphasized the importance of the radial routes outwards from the centre of cities and showed the way in which the high status areas, which in his view determined much of the other patterning within the city, tended to move outwards along these radial lines in distinct sectors of the city. In terms of general principles 'the movement of the high-rent area is in a certain sense the most important since it tends to pull the growth of the entire city in the same direction'.[3] The high status areas themselves, once established in a certain sector of a city, would tend to grow or expand outwards within that sector. Other sectors which began to develop as low-rent residential sectors similarly retained the same character for long distances outwards as the low-rent housing extended with the process of urban expansion. Giving prime importance to the high-rent sector, Hoyt then suggested that the point of origin of this sector was determined by the location of the retail and office centre where the higher income population tended to work. Starting from close to this centre its movement was determined by a number of considerations: growth proceeded along lines of travel or towards another nucleus of buildings or trading centres; it progressed towards high ground free from flood risk and along bay and waterfront areas, where such areas were not pre-empted by industrial development; growth

---

[1] M. R. Davie, 'The pattern of urban growth', in G. P. Murdock (ed.), *Studies in the science of Society* (New Haven, 1938), pp. 133–61.

[2] *Ibid.* p. 161.

[3] H. Hoyt, *The Structure and Growth of Residential Neighbourhoods in American cities* (Washington, D.C., 1939), p. 114.

occurred towards areas of open country avoiding 'dead end' sections. Such are some of the considerations concerning the paths of movement of the high-rent sector which in turn controlled other types of residential areas. Allowance was also made for the influence of real-estate promoters and an exception to the general outward movement was that high status *de luxe* apartments tended to be built close to the central parts.

Given the development of segregation of types of land occupance within a growing central business district, Hoyt's scheme thus suggests the principles from which it should be possible to predict the future lines of residential development. The essential difference between his very flexible model and that of Burgess is that Hoyt added a directional component to the evaluation of the spatial patterning of urban residential areas. Nevertheless, he still recognized the validity and importance of the centrifugal/centripetal forces upon which the Burgess model was based. For example, in considering the sequent changes within the city, he stressed the fact that change of occupance was essentially centrifugal as the city expanded. As houses became older and more deteriorated, invasion of different succeeding populations (to use the terminology of the classical ecologists) occurred. However, within this centrifugal pattern he distinguished a directional component which derived from the different types of houses which had been vacated by the various income groups. High status housing, for example, was subject to an 'extraordinary rate of obsolescence', since, when the wealthy left such areas, there was no class of people willing or able to occupy single houses because the expense of upkeep was only within the range of the wealthy. No class filtered in to occupy the houses for single-family use and they were thus converted into boarding houses, offices, clubs or light industrial use. By comparison, houses within the medium-rent sectors were taken over by families with slightly lower incomes as the houses lost some of their original desirability. There was thus some loss of value but no essential change in the type of use of such sectors. Finally, the low-rent sectors underwent great change as they deteriorated since the worst buildings were demolished and, unless subsequent waves of poor immigrants continued to enter the city and provide a demand, the areas of submarginal housing were removed. Thus, in Hoyt's scheme, as in Burgess', 'the erection of new buildings on the periphery of a city, made accessible by new circulatory systems, sets in motion forces tending to draw population from the older houses and to cause all groups to move up a step leaving the oldest and cheapest houses to be occupied by the poorest families or to be vacated'.[1] The difference, however, was that Hoyt suggested that this invasion, and therefore the types of social area, differed from sector to sector. The rooming-house area, for example, was found most typically at

---

[1] H. Hoyt, p. 122.

the central apex of the high status sector since it was here that large old houses provided conditions ripe for subdivision and further decay.

While Hoyt's scheme was thus radically different from Burgess' in many ways, the dynamic mechanisms which it isolated are perhaps much less different in essence than many commentators have tended to suggest. While recognizing the importance of outward centrifugal movement he added a directional element which distinguished *sectors* of growth and segregation in addition to the concentric *rings* which are the chief feature of the Burgess model.

In addition to such overt or implicit criticisms of the overall model of urban development, there were specialized empirical studies which were aimed at evaluating and criticizing aspects of the ecologists' views. Paul Hatt, for example, criticized the concept of the natural area: from studies in Seattle, Hatt suggested that, rather than the city being composed of a number of homogeneous natural areas, which were 'real' and which acted as coercive influences upon all who dwell in them, different sets of data—rental, ethnic, and social—produced different sets of areas whose boundaries were blurred and were usually overlapping. He suggested that the ecologists tended to reify the concept of the natural area and that instead 'ecological phenomena [should be] viewed as a frame of reference and the concepts of natural areas as a construct within that frame'.[1] Only thus could his data be assigned to natural areas—each different, depending on the type of data used.

Perhaps the most fundamental, and the most widely known, of the empirical criticisms was Firey's work on Boston.[2] Firey emphasized the sentimental and symbolic connotations of area and place and illustrated the manner in which such factors could counteract the ecological sub-social forces which were based on competition and the 'rational' allocation of land uses. He used three areas to substantiate his thesis. The first was the exclusive inner area of Beacon Hill. Unlike other inner high status residential areas which, as ecological theory would predict, suffered decay, invasion and consequent loss of status over time, Beacon Hill maintained its exclusive character over a period of more than half a century because of the sentiment and symbolism associated with its aesthetic, literary, historical and familial traditions. Secondly, the traditional burial grounds, common land and the old colonial churches within the inner areas were preserved in spite of commercial and business pressures for expansion which would engulf them. Finally, the North End, an area of dilapidated slum dwellings, which the classical ecologists would view as an area of minimum choice, held certain positive

---

[1] P. Hatt, 'The concept of natural area', *Am. Sociol. Rev.* XI (1946), 426.
[2] W. Firey, 'Sentiment and symbolism as ecological variables', *Am. Sociol. Rev.* X (1945), 140–8; *idem, Land use in central Boston* (Cambridge, Mass., 1947).

retentive values for the first-generation Italians who lived in the area. In essence Firey's argument was that sentimental attachments and symbolic associations could become identified with certain areas and that they could effectively countervail against the biotic forces stressed by the ecologists. Man's culture and the mesh of social stimuli surrounding his actions and decisions could not be ignored or thought away in an attempt to carry through a biological analogy to account for the structuring and patterning of towns.

Theoretical criticism

It was this latter point which was most strongly attacked in the theoretical criticism of the period. Milla Alihan's book of 1938 was the most effective and persuasive theoretical criticism of the ecological position.[1] In it she presented a careful critique of the concepts underlying the ecologists' viewpoint. She pointed out such inconsistencies as the belief in both natural areas and gradients, which she suggested were mutually exclusive concepts. She criticized the ecologists' excessive dependence on the tangible: 'the human ecologists undertook to explain the social complex by fastening attention upon its salient manifestations, such as the growth of cities, the spread of industry, the extension of railways and highways, the mosaic of nationalities and races, the movement and distribution of people and utilities. The conditions of social change became to them *facts* of social change...and their interpretation of social life hinged upon its most concrete aspects.'[2] The main attack was levelled at the haziness of the ecologists' theoretical formulations and in particular at their fundamental distinction between the biotic 'community' and the cultural 'society'. Alihan argued that such a distinction was in any case one which could not be upheld. The processes involved in the dichotomy were so highly intertwined that their treatment invariably resulted in their fusion. The ecologists 'find that the presence of "group economy" which defines a plant or an animal colony, is in human community so highly evolved and so intrinsically a part of other social phenomena that the analogy becomes worthless and the ecologists are forced to define the very concepts intended to describe this organic aspect of life in terms of the assumed social-psychological concept "society". Consequently, what is said to be unsocial in one instance is asserted to be social in another.'[3] Furthermore, empirical studies made by ecologists did not conform to this ill-conceived and loosely delineated theoretical distinction. She argued that, in their monographs and articles, the ecologists included social elements within studies which ostensibly were concerned with organic or biotic aspects of urban structure. Very

[1] M. A. Alihan, *Social ecology : a critical analysis* (New York, 1938, reprinted 1964).
[2] *Ibid.* p. 6.   [3] *Ibid.* p. 85.

rarely was it made clear what the biotic aspects were supposed to represent. Indeed, she dismissed these works as being, not ecological as defined by its theory, but rather as being 'general sociological studies in which territorial distribution is taken account of in reference to sociological data'.[1] Thrasher's study of the gang was a case in point since, while being ostensibly ecological, he studied the gang as a social grouping exhibiting a certain type of behaviour, and the only ecological aspect which was introduced was the delineation of the territorial area in which the gang was found. His concern was therefore not restricted to the so-called biotic elements, but rode roughshod over the theoretical framework to include both social and sub-social aspects.

Alihan's criticism of the community/society dichotomy was also the main target of later criticisms. Gettys, for example, concentrated on the indivisibility of the social and ecological aspects of urban structure, arguing like Alihan that decisions and behaviour could not be separated into planned and unplanned, or biotic and cultural dimensions. Such 'dichotomies...tend to disappear when it is recognized that culture, in its varied and multiforms, wherever found, together with human motives of great variety, influence, condition, and even in some instances, determine where and by what means people shall live; that so-called symbiotic relations among human beings are not immune to the influence of choice, taste, initiative, desire, and customary and institutional controls; that competition is not free and unrestrained, particularly among more civilized peoples; that the equilibrium, designated as "biotic", is very much a matter of agricultural and industrial methods, technological achievements, systems of economic distribution, dietary practices, sanitation, development and application of medical knowledge, conscious family limitation, and war-making tendencies.'[2] Gettys thus makes the same point that Firey made in his empirical studies of Boston; that culture forms a matrix which modifies man's evaluation and exploitation of his environment and which cannot be ignored or thought away by the process of dichotomizing. He also criticized the ecologists for the determinist slant of their approach to human behaviour, a bias which was made almost inevitable by their strong dependence on analogies drawn from the organic ecologists.

In criticizing the ecologists, these writers also suggest ways out of the dilemma which arises from faulty concepts. Alihan, noting that much of the ecological writings concentrated upon the tangible external factors of human behaviour such as the distribution and movement of population, utilities and physical structures, argues that, while there was no objection to this abstraction of external phenomena, they were not the intrinsic factors of the

---

[1] *Ibid.* p. 83.
[2] W. E. Gettys, 'Human ecology and social theory', *Social Forces*, XVIII (1940), 471.

community as defined by ecological theory. She suggested that they should abandon their theoretical concepts and concentrate upon the development of a subject which studied the 'civilizational' aspects of society or one which only studied spatial distribution. The discipline would then be delineated either by its subject matter or by its specific approach. Gettys, concerned more with the relations of human ecology to sociology, suggested the need for more precise delimitation of the field of human ecology and its place *vis-à-vis* social science as against biological ecology. He suggested that it should abandon its dependence on the natural sciences and centre its attention upon the 'description, measurement, analysis, and explanation of the spatial and temporal distributions of social and cultural data'. Only then could it become a significant social science discipline.

Reaction to criticism

After more than a decade of such empirical and theoretical criticism, Foley was able to comment: 'by now the sociologist's intellectual honeymoon with urban ecology is over and he is faced with the problem of settling down and living with this ecological approach'.[1] The mounting criticism resulted in re-orientations of varying kinds as well as a general loss of intellectual confidence amongst ecological writings even though they continued unabated. The main theoretical readjustment was prompted by the need to incorporate certain aspects of the cultural facets of urban society into the ecological approach. Theodorson makes a rather unsatisfactory distinction between a neo-orthodox and a socio-cultural approach amongst the ecological studies which followed the wave of criticism.[2] The distinction is a rather loose one, of degree not of kind, since it distinguishes those ecologists who gave greater or lesser primacy to the importance of cultural factors in the explanation of 'ecological' structures. The more important general point is that ecological writing began explicitly to take great cognizance of the role of culture. It is significant that it was during this period of re-orientation that the first major attempts to delineate the field of human ecology in a systematic fashion appeared. Previously, in spite of the wealth of literature, such an attempt had not been made and indeed one has to perform mental gymnastics to gather together the elements of Park's theoretical doctrine which are scattered throughout his writings.

Two major theoretical statements, however, emerged in the period following the criticism of the 1930s. The first was by Quinn, whose stated position differs but little from that of the classical ecologists.[3] His distinction

---

[1] D. L. Foley, 'Census tracts and urban research', *J. Am. Statist. Ass.* XLVIII (1953), 736.
[2] Theodorson, *Studies in human ecology.*
[3] J. A. Quinn, *Human Ecology* (New York, 1950).

20

between a cultural and a sub-cultural level of society derives directly from the classical dichotomy of cultural/biotic. It was Hawley's statement of the scope of the subject which has been taken as the starting point for more recent approaches to ecology.[1] His approach is indicated in the subtitle of his work: 'a theory of community structure'. 'Community' is seen as the more familiar name for ecological organization as applied to a unit of ecological investigation. 'Organization' in turn describes the way in which a variety of discrete phenomena exist so as to constitute a larger unit. Such units are organizations only if their component parts perform one or more functions: for example, a rock would not be considered an organization whereas a motor car would. Hawley's definition thus places great stress on the concepts of functional relations in the adjustment of human aggregates to their organic and inorganic surroundings. This emphasis on functional organization differs from earlier definitions which stressed the importance of the spatial patterns which such functions assumed. In considering that the realm of human ecology is the study of the form and functioning of the community and of the development of the community, Hawley discusses the nature of ecological organization. This he sees in terms of three aspects: differentiation, community structure and spatial structure.

Differentiation is the basis of ecological organization. Without differentiation there cannot be organization and this differentiation is based not only on what Hawley calls 'physio-psychological' differences—age, sex and race factors—but also on territorial factors. His analysis places a great deal of weight on the fact of the division of labour and may therefore be viewed as basically economic in its orientation.

Community structure, based on differentiation, thus forms the focus of ecological study, and Hawley here draws on certain of the terms of the classical ecologists to develop a typology of communities. In place of Durkheim's dichotomy in time between 'mechanical solidarity' and 'organic solidarity', or of Tönnies' dichotomy in place between *gemeinschaft* and *gesellschaft*, Hawley evolves an elaborate and confusing typology involving symbiotic relations on the one hand and commensalistic relations on the other. Symbiotic relations relate to the interdependence of unlike forms, units of dissimilar functions, drawing on the analogy of such symbiotic relations in the animal world as the interdependence of, say, the carnivores and the herbivores or the leech-eating plover and the crocodile. Commensalistic relations, on the other hand, relate to the interdependence between like forms, units of similar functions, and are expressed in terms of competition. By combining the two 'it appears, in fact, that the collective life of man, as of all other organisms, revolves simultaneously about two axes, one

[1] A. H. Hawley, *Human Ecology: a theory of community structure* (New York, 1950).

of which is symbiotic, the other commensalistic'.[1] The functions and forms of units can thus be studied in terms of the axis to which the unit is assigned. Communal units based on symbiosis are 'corporate units'; those based on commensalism are 'categoric units'. Corporate units are those linking dissimilar types of function and include such organizations as familial, associational and territorial units. Categoric units, linking similar types of function, are largely occupational but can also be based on neighbourhood units, clubs or cliques of various sorts. The distinction between these groupings is highly unsatisfactory since, even by Hawley's admission, there is a great deal of overlap and confusion as to which parameters are being used to identify whether the grouping is of like or unlike functions. For example, with a change of social situation, the categories break down. 'Thus', as Hawley says to add to the confusion, 'the corporate unit may function, particularly when its unity is threatened, as a categoric unit. And the categoric unit may develop a corps of specialists—a symbiotic nucleus within the core of commensalists.'[2] The analysis of communities is then continued with a dichotomy drawn between the 'dependent' and the 'independent' communities, which appear to be similar to Durkheim's categories of mechanical and organic solidarity, but expressed in ecological terminology. Here, Hawley is perhaps at his weakest in his attempt to incorporate elements of the terminology of the organic ecologists, and thus to sustain his ecological view of society.

Having examined the role of differentiation and community structure in ecological organization, he goes on to consider the role of spatial structure—the third element of ecological organization. While stressing the subordinate role that spatial analysis plays in his view of human ecology, he considers that spatial distribution is important in so far as it reflects aspects of functional interrelationships. There is a spatial aspect to community form and function even though that spatial structure does not incorporate all the functional elements of interest to the human ecologist. In looking at these spatial aspects, he is concerned to give spatial expression to the concept of 'community' and distinguishes between the independent community whose boundaries are frequently precise and clear, and the dependent community where the boundaries are much less clearly demarcated. In the more evolved communities, in other words, he recognizes that 'the organizations of spatially separate populations...have merged at many points giving rise to very extensive and inclusive communities. The term *community*, interpreted to connote a compact, easily distinguishable entity, has lost much of its meaning.'[3] He thus introduces the factor of recurring movements to define the community in terms of 'that area the resident population of which is interrelated and integrated with reference to its daily requirements, whether contacts be

---

[1] Hawley, *Human Ecology*, p. 209.   [2] *Ibid.* p. 232.   [3] *Ibid.* p. 255.

direct or indirect'.[1] In the complex dependent community such an area would therefore include urban areas plus their primary areas (of daily circulation of population) and secondary areas (weekly and less frequent trips). In considering the spatial structure of the community on a more detailed scale, the focus of the study seems to change. While the foregoing theoretical considerations have dealt with human aggregates and social groups, the analysis of the spatial structure of the urban area is largely in terms of such external phenomena as land-use features. Instead of considering his corporate and categoric units, the elements of study are manufacturing establishments, retail store distribution and so forth.

In all, Hawley's statement of the scope of human ecology is the most carefully developed and most authoritative and yet at the same time unsatisfactory in that it contains a number of ill-enumerated concepts. The basic criticism which was levelled at the human ecologists of the 1920s and 1930s—that the cultural aspects of human interaction and behaviour could not be isolated or thought away—has been met by providing a theoretical system which accepts that all human interrelationships are social, and by concentrating on aspects of organization which have an economic basis. Yet the dividing line between cultural and non-cultural aspects has not been satisfactorily demarcated. For example, in noting Firey's criticisms, that ecology makes the assumption that the basis for human behaviour is rational, Hawley notes the similarity on this point between ecology and economics and comments: 'Ecologists are as fully aware of the force of habit and sentiment as anybody else. As a matter of fact, however, the issue of rationality versus irrationality does not concern us. Human ecology studies the structure of organized activity without respect to the motivations or attitudes of the acting agents. Its aim is to develop a description of the morphology or form of collective life under varying external conditions. With its problem stated in that manner the irrelevance of the psychological properties of individuals is self-evident.'[2] Yet the content of his work and the very subtitle of the volume show that Hawley's view of ecology is not as a descriptive discipline but as a developed theory of the community. A viewpoint which emphasizes the *functional* role of social institutions, as does Hawley's, makes assumptions as to motivations, attitudes, sentiments and values which must at least be recognized and considered. Yet Hawley argues: 'Attitudes, sentiments, motivations, and the like are omitted from consideration not because they are unimportant, but because the assumptions and point of view of human ecology are not adapted to their treatment.'[3] Such a disclaimer, however, does not avoid the necessity to clarify the interrelations between human ecology and such psychological factors. In thus assuming that institutions

[1] *Ibid.* pp. 257–8.   [2] *Ibid.* p. 179.   [3] *Ibid.* p. 180.

have a function to play in the process of the adaptation of human groups to their total environment, but by failing to clarify the social context within which such functional interrelationships might be assumed, Hawley has merely sidestepped most of the criticism which was levelled at the ecologists of the Chicago school.

The modern theorists of human ecology have used Hawley as their springboard. In discussing the scope of the subject, Duncan for example begins by isolating a recurring theme of human ecology, the relationship between men and their environment, and then comments that Hawley's theoretical exposition represents a consistent well-elaborated version of the discipline.[1] He then proceeds to build upon Hawley's work to formulate a 'theory' of human ecology as a frame of reference. It begins with a unit of observation: a human population more or less circumscribed territorially. This population exists within an environment, and between the two there is action and reaction in the process of an adjustment which is continuous and dynamic. This problem of adjustment is both complicated and facilitated in a human group by man's possession of culture and, within the realm of culture, social organization and technology play important parts either as dependent or independent variables within the process. There are thus four 'referential concepts' of the main problems of human ecology: population, environment, technology and organization. Duncan suggests the scope of the ecological perspective by linking the four as four vertices each of which is linked by lines which suggest their 'functional interdependence', and the whole system is labelled 'the ecological complex'. This provides a framework for ecological research, and the main tenets of its approach are that society exists by means of the organization of a human population whose members are individually unequipped to survive in isolation. 'Organization represents an adaptation to the unavoidable circumstance that individuals are interdependent and the collectivity of individuals must cope with concrete environmental conditions —including, perhaps, competition and resistance afforded by other collectivities—with whatever technological means may be at its disposal. The "social bond", its most basic aspect, is precisely this interdependence of units in a more or less elaborated division of labour, aptly described as a "functional interdependence".'[2] Schnore has written in a similar vein demonstrating the parallels between this 'neo-ecology' and the social morphology of Durkheim. The closeness of the theoretical positions of Duncan and Schnore has been crystallized in a joint methodological paper.[3]

The role of culture in this methodology is more clearly stated. Duncan

[1] O. D. Duncan, 'Human ecology and population studies'.   [2] *Ibid.* p. 683.
[3] See L. F. Schnore, 'Social morphology and human ecology'. *Am. J. Sociol.* LXIII (1958), 620–34; and O. D. Duncan and L. F. Schnore, 'Cultural, behavioural and ecological perspectives in the study of social organization', *Am. J. Sociol.* LXV (1959), 132–46.

24

denies that the ecologist constructs a biologistic theory of human behaviour which ignores cultural factors, but that the human ecologist finds the concept of culture too global and synthetic to enable it to deal with the system of interdependencies which fall within the ecological complex, so that 'the functional and analytic approach of human ecology involves a concern not with culture as an undifferentiated totality but with aspects of culture as they play into the process of adaptation'.[1] Social organization, which is presented as a rather ill-defined concept, and technology are the two components which presumably incorporate these aspects of culture, but values are explicitly excluded from their treatment.

Certainly the formulations of Duncan and Schnore have provided a much more rigorous framework within which ecological problems can be slotted, but in providing the framework they have perhaps been guilty of what Rossi calls 'a distressing tendency toward intellectual "imperialism"'. As he says, 'The ecological perspective is so loosely defined that it can be stretched to include what is praiseworthy...and contracted to avoid the apparently faulty'.[2] Nor have they satisfactorily taken account of the importance of values and the full range of cultural variables as factors within the process of ecological adaptation. As Sjoberg notes, 'the urban field is a major battlefield for those who stress the impact on urban life of "objective conditions"—the external environment, population structure and the like—and those who emphasize, for instance, the role of social or cultural values as a key determinant of the so-called objective conditions and of human action in general'.[3] This has been a recurring criticism of all ecological work and one to which we shall return later.

The empirical work of Duncan and Schnore has, however, added greatly to our knowledge of urban structure and of the functioning of urban communities. Duncan, for example, has studied aspects of the urban hierarchy in American cities, thus following on from some of the early work of McKenzie. Many of his works on individual urban areas have made stimulating contributions to the analysis of the spatial interrelationships of urban phenomena.

RELATIONS BETWEEN GEOGRAPHY AND ECOLOGY

Consideration of the history of the development of ecology and of geography in their approaches to urban problems suggests a great many parallels between them. To clarify this relationship it will be profitable to examine

---

[1] Duncan, 'Human ecology and population studies', p. 682.
[2] P. H. Rossi, 'Comment', *Am. J. Sociol.* LXV (1959), 148.
[3] G. Sjoberg, 'Theory and research in urban sociology', in P. M. Hauser and L. F. Schnore (eds.), *The study of urbanization* (New York, 1965), p. 159.

more closely some of the overlaps and similarities. This will be done in terms of what the two subjects have actually studied, what their practitioners have considered to be the relationship between the two and, finally, what a pragmatic consideration of the nature of the two subjects suggests.

Concepts and fields of study

Certainly many of the key concepts of ecology and human geography have been remarkably similar. For example, the idea of invasion and succession which was first applied by the Chicago ecologists is mirrored in geography by the concept of sequent occupance which played a large part in the analyses of such geographers as Sauer. The fact that the one concept was applied to towns while the other was applied to larger areas or 'regions' is a reflection of the preoccupation of geographers in the 1920s with rural areas to the neglect of towns.

Similarly, the ecological concept of the natural area finds its geographical counterpart in the natural region. Both were seen as areas of homogeneous social and physical characteristics and, in the early years, both were seen as being the result of deterministic forces which moulded the phenomena found within them. Burgess' comments about the natural area might be applied directly to much of the geographical thinking concerning the natural region, the only difference being one of scale and the fact that, again, one was urban the other was not: 'This differentiation into natural economic and cultural groupings gives form and character to the city. For segregation offers the group, and thereby the individuals who compose the group, a place and role in the organization of city life. Segregation limits development in certain directions, but releases it in others. These areas tend to accentuate certain traits, to attract and develop their kinds of individuals, and so to become further differentiated.'[1] The controversy surrounding the concept of the ecologists' natural areas is likewise reflected in the geographical rethinking of the *pays* and of the regional approach of such men as Vidal de la Blache.

Yet another point of similarity is found between the concept of the community, as delineated by Hawley, and the geographer's conception of a functional region. Actually, in considering the community region, Hawley ignores these parallels. He contrasts what he calls the 'ecological community areas' of the United States with so-called 'geographic regions' and concludes that the community region is a different kind of spatial unit from that delineated by the geographers.[2] The community is defined in terms of areal interdependence, which is illustrated by reference to the circulation of news-

[1] Burgess, *Publ. Am. Sociol. Soc.* XVIII (1924), 56.
[2] Hawley, *Human Ecology*, pp. 258–62.

papers: the geographic region by contrast is defined in terms of homogeneity of economic characteristics. Yet this limited interpretation of geographical regions ignores the fact that geographers, like ecologists, have long recognized a type of region which is delineated in terms of areal interdependence. As early as 1935, well before the appearance of Hawley's work, the geographer Platt could draw a distinction between two general regional concepts in geography which related to 'static areal homogeneity' on the one hand and to 'areal functional unity' on the other.[1] By 1947, drawing on work done in the 1930s, Dickinson contrasted the homogeneous region with the functional region, defining the latter as 'an area of interrelated activities, kindred interests and common organizations, brought into being through the medium of the routes which bind it to the urban centre'.[2] This distinction within geography has developed into the concepts of the uniform and nodal (or, alternatively, the formal and functional) regions; the one defined in terms of overall homogeneity, the other in terms of functional interconnection in economic or social respects. Hawley's disregard for this well-established concept is therefore somewhat difficult to understand.

Like the ecologists' 'community', the geographical functional region is delineated with respect to patterns of circulation such as the journey to work, or the movement of goods, or the supply of services; the very parameters which Hawley uses to delimit the ecological community. The geographer's functional region, primarily the concept of the central city and its dependent region, is identical in virtually all respects to the ecologist's concept of community.

With this shared battery of basic concepts, it is not surprising that many of the topics studied by geographers and ecologists should overlap. In view of the similarity between functional regions and communities, for example, it is natural that the study of the metropolitan region should be found in both subjects. The nature and workings of the connection between a central place and its tributary area have long been a feature of both disciplines. In geography the voluminous literature on spheres of influence is in effect no different from the ecologist's concern with functional interdependence in area. Indeed, a summary of the topics included within ecological and geographical text-books as falling within the scope of the respective subjects shows a large number of such overlaps. Duncan, for example, in his article on ecology and demography suggests the scope of the ecological treatment of demographic problems.[3] First is the study of population distribution, in which consideration is given to frequency distributions of town size and a

[1] R. S. Platt, 'Field approach to regions', *Ann. Ass. Am. Geogr.* xxv (1935), 171.
[2] R. E. Dickinson, *City region and regionalism* (London, 1947), p. 11.
[3] Duncan, 'Human ecology and population studies'.

discussion of some of the empirical regularities observable within such distributions. Included here are discussions of Zipf's rank-size rule and Stewart's extensions of it. Another aspect of population distribution which is considered is the spatial pattern of towns and cities with discussion of locational analysis and the spacing of central places. A final aspect is the distribution of density gradients. Duncan suggests that human ecologists use such distributional data where the data serve as indicators of aspects of ecological organization, but it is interesting to note that all of the topics considered as falling within the ecological sphere are also included by Berry as being among the 'major substantive topics' considered in the research frontiers of urban geography.[1] Berry, like Duncan, discusses systems of cities, models of city-size distributions and density gradients and calls on the same authorities and the same references as does Duncan. The other ecological aspects of demography which Duncan considers also have counterparts in the geographical literature. He looks at population composition, the vital processes, movement and population growth, and in his summary of problems and issues he isolates 'the environment and regional analysis' from the population balance. In urban geography, economic geography and population geography the same issues and the same treatment can be found. Indeed, Ackerman notes, 'The common ground of human ecology and human geography is suggested by Duncan's reference to the works of ten different geographers in the presentation of his chapter'.[2] Schnore, too, has himself noted the extent of the overlap between the subject matter and the concerns of the two disciples. Again, the work of Berry is noted and Schnore compares his own treatment of economic development and urbanization with that of Berry. He concludes: 'these two independent efforts are practically identical in approach and results, despite the different disciplinary identification of their nature'.[3] He also suggests the similarities between two basic collections of writings; one on urban geography, the other on urban sociology. Again the conclusion is that they 'reveal markedly similar preoccupations'. Further, the geographical studies of the urban economic base 'are as structural in emphasis as anything in the ecological literature'.

The views of practitioners

This therefore gives some idea of the overlap in content and in interest between the works of ecologists and geographers. In view of the similarity

[1] B. J. L. Berry, 'Research frontiers in urban geography', in Hauser and Schnore (eds.), *The Study of urbanization*, p. 403.
[2] E. A. Ackerman, 'Geography and demography' in Hauser and Duncan (eds.), *The study of population*, p. 723.
[3] L. F. Schnore, 'Geography and human ecology', *Economic Geography*, XXXVII (1961), 215.

it would be surprising if writings on the methodology of the subjects had not called attention to the degree of commonness. The earliest such methodological writing is Barrows' plea that geography should be considered as human ecology, written at the outset of the work of the classical ecologists.[1] Barrows' suggestion for 'humanizing' physical geography was a modification of the environmentalist concept. By studying the 'mutual relations between man and his natural environment' geography would become the science of human ecology. These relationships should be studied, however, 'from the standpoint of man's adjustment to environment, rather than from that of environmental influence'. This avoids 'the danger of assigning to the environmental factors a determinative influence which they do not exert'. The fact that his proposals elicited little response from American geography may have been a reflection of the fact that the qualitative and quantitative techniques of the time were unable to exploit the full value of the ecological concept.[2] It may also be that the exclusion of physical geography itself, which was the desired consequence of Barrow's methodology, led to the neglect of his appraisal. Nevertheless, the recurring theme of geography as a study of man/land relations inevitably led to a number of geographers viewing the subject as being ecologically orientated. White and Renner, for example, subtitled their geographical work 'an ecological study of society'[3] and it was in terms of this concept of the relations of man to his natural environment, rather than in terms of the more fully articulated theoretical framework of the human ecologists, that this subtitle is to be interpreted.

A second strand in the ecological view of geographers is seen in Bews' work on human ecology which sought to bring a great deal of geographical work under the umbrella heading of human ecology. Bews' approach was essentially holistic since he viewed the subject as 'not so much a branch of science as a certain attitude of mind with regard to life', or again, 'while in one sense, ecology is merely a point of view, in another sense it is the most complete science of life, since life is not a thing itself but a *process*, which of necessity continuously involves the environment'.[4] This holistic stress on man and his environment is seen too in his emphasis on an approach which stresses the 'environment, function, organism triad'. That his view of human ecology was little different from the geography of the day is obvious from a review of his chapter headings which include consideration of the environment, heredity, response to the environment, control of the environ-

[1] H. H. Barrows, 'Geography as human ecology', *Ann. Ass. Am. Geogr.* XIII (1923), 1–14.
[2] A view expressed by E. A. Ackerman, 'Where is a research frontier?', *Ann. Ass. Am. Geogr.* LIII (1963), 431 n.
[3] C. L. White and G. T. Renner, *Human geography : an ecological study of society* (New York, 1948).
[4] J. W. Bews, *Human Ecology* (London, 1935), pp. 1 and 278. This view of ecology properly lies within the scope of 'general ecology'—the fourth element of ecology as defined above, p. 8.

ment and a review of man/environment relations at different stages of economic development. In his discussion of 'ecological studies' he considers the regional surveys of such scholars as Patrick Geddes and Le Play and the social surveys of York, Tyneside, Merseyside and elsewhere.

This macroscopic or global view of ecology, which studies man and his relations with the environment, is a realm within which geography and 'general ecology' have been consciously linked by a number of writers. More recently, it is a theme which has been restated in terms of new ecological concepts and the ideas of general systems theory. The students who have traced the links of the two subjects in these terms have been orientated towards the physical rather than the human aspects of geography and have presented the parallels between the subjects in terms of the organic concepts of the ecosystem. The most persuasive of these writings is by Stoddart, who relates systems theory to a quantified approach to environmental studies.[1] These suggestions within geography, however, are moving from the organic ecological complex back into the physical realms and, even though human activities are included within the purview of consideration, they are essentially a synthesizing extension of plant and animal ecology which includes the wide range of phenomena falling within the ecosystem. Their relation to *human* ecology and human geography is tenuous by the nature of their prime concern.

While certain geographers have thus stressed the connections between geography and ecology, although seeing both subjects in various guises, most ecologists have consciously tried to differentiate the two disciplines. Their attempts have not always been very convincing. Hawley, writing in 1950 and drawing upon geographical sources which are in no case more recent than 1936, views geography as a subject which is concerned only with the relationships between man and his 'natural' environment; and with this rather inadequate concept of geography he draws two distinctions between it and human ecology. First, as geography is preoccupied with the physical environment and man is only introduced as 'he is a part of the natural landscape', it cannot concern itself with the interrelations among men; human ecology is primarily interested in the interdependencies which develop in the action and reaction of a population to its habitat. Thus, 'while geography views the adjustment of men from the standpoint of modifications of the earth's surface, human ecology makes a detailed analysis of the process and organization of relations involved in adjustment to environment'.[2] Secondly,

---

[1] D. S. Stoddart, 'Geography and the ecological approach', *Geography*, L (1965), 242–51. For somewhat similar views see S. R. Eyre, 'Determinism and the ecological approach to geography', *Geography*, XLIX (1964), 369–76, and W. B. Morgan and R. P. Moss, 'Geography and ecology: the concept of the community and its relationship to environment', *Ann. Ass. Am. Geogr.* LV (1965), 339–50.   [2] Hawley, *Human Ecology*, p. 71.

whereas geography involves a static approach, viewing things as they are at a given point of time, ecology is dynamic and evolutionary, describing the developmental process as well as the form of human adjustment. The first point in particular is hard to defend. Even at the time that Hawley was writing, much geographical work was dealing with the relations of man to man. To say that geographers have studied only man/environment or that such a view would encompass the scope of the subject is to ignore a great deal of what has been most significant within its writings and methodological development. It is true that, at the time of Hawley's writing, fewer geographers were concerned with the interrelationships of human characteristics than is the case today, yet this false basis for differentiating the two subjects has been one which has consistently appeared despite much contrary evidence in the geographical literature. As late as 1961, for example, Schnore repeated the ideas of both Hawley and Erickson, that geography and ecology could be distinguished by the fact of ecology's interest in the relations between man and man as influenced, among other things, by his habitat, as against geography's more limited concern with man's relationship with his environment.[1] While there may have been some element of truth in such a view at the time of Hawley's book, by the 1960s it could certainly not be held that geography ignored the relations between sets of phenomena which were wholly human. In fact, it was only three years after Hawley's work that Trewartha was writing to urge the study of population geography. His 'trinitarian approach', which replaced the old dualism between physical and human phenomena, comprised man, the natural earth and the cultural earth in a system which is very similar to Duncan's ecological complex. Trewartha's belief was that 'fundamentally, geography is anthropocentric and, if such is the case...numbers, densities and qualities of the population provide the essential background for all geography. Population is the point of reference from which all the other elements are observed and from which they all, single and collectively, derive significance and meaning. It is population which furnishes the focus.'[2] Such glowing stuff was not only a reflection of an existing and growing dissatisfaction with the old man/environment approach, but also reflected the increasing emphasis which was being given, and has been given increasingly, to the geographical treatment of the interrelations of purely human phenomena. In terms of theoretical orientation, much the same point can be made from an examination of a large number of statements of human geography. Wrigley, for example, argues that the old

[1] Schnore, *Econ. Geogr.* XXXVII (1961); E. G. Erickson, *Introduction to human ecology* (Los Angeles, 1949). While Schnore does add footnotes and riders to decry some of the views which he quotes, his chief purpose appears to be to 'underscore the essential differences between the two subjects'.

[2] G. T. Trewartha, 'The case for population geography', *Ann. Ass. Am. Geogr.* XLIII (1953), 83.

formula of Le Play should be turned on its head so that geographical study would start from man, develop to include economic organization and finally include relevant environmental aspects.[1]

Another paper by an ecologist which looks at the relations of the two disciplines and concludes by 'underscoring their essential differences' is Schnore's which, as has already been noted, accepts the very great degree of overlap in their empirical interests. In spite of the tenor of the remainder of the paper which illustrates the virtual indistinguishability of the two, he concludes that the historical contrast is most enlightening since, 'as geographers have come to realize that the core of their interest is the study of *areal and spatial differentiation*, ecologists have simultaneously turned away from a concern with spatial patterns *per se*, except as they enlighten us with respect to *social organization*'.[2] Such a conclusion may be supported by selective culling of the theoretical writings of the two disciplines if one wished to emphasize the differences of the two subjects, but a very different conclusion could be argued in theoretical terms. Certainly an examination of the empirical writings and approaches of the subjects does not support such a conclusion, as Schnore himself admirably demonstrates.

An inductive approach to definition

An attempt to examine the relations between the two subjects from a study of the theoretical writings of its practitioners is subject to numerous difficulties. The opinions of selected geographers may not adequately reflect the currently accepted views of other geographers, and there is also the difficulty that, since both subjects have been in a state of considerable flux throughout this century, the views of the relations between them are peculiarly time-bound. It may have been that at any given time one subject has stressed one particular aspect which has only gained prominence in the other subject at some later date. An alternative approach is thus the more pragmatic one of trying to isolate the essential features of each discipline and to compare them irrespective of the theoretical stances of the practitioners.

It has been noted already how great is the overlap in the substantive contents of human ecology and human geography. Two elements above all have been recurring themes in each: on the one hand an interest in space and on the other an interest in adaptive processes. It has been argued that geography is essentially a correlative discipline interested in the spatial associations of phenomena. From the first this was also a concern of ecology. In the Chicago school there was an unashamed interest in space almost for its

[1] E. A. Wrigley, 'Geography and population', in R. J. Chorley and P. Haggett (eds.), *Frontiers in geographical teaching* (London, 1965), pp. 62–80.
[2] Schnore, *Econ. Geogr.* XXXVII (1961), 217.

own sake. The more recent proponents of the ecological approach have tried to pare down this spatial content so that spatial aspects become significant only in so far as they are a reflection of the functional relationships and organization of human aggregates. Yet, as Duncan says, 'that at least some spatially delimited population aggregates have unit character is one of the key assumptions of human ecology'.[1] The concern with place and with location is seen throughout ecological studies. But does the caveat which ecologists add to this interest make it in any way different from the geographical approach to space? Geographers, no less than ecologists, are not interested in spatial patterns *per se* but rather in the dynamics and the factors which are responsible for the existence of spatial associations or spatial patterns. As Ackerman has said, 'geographers who have thought in terms of areal functional organization...have had a significant insight as to research direction'.[2] True, the geographer starts out with the identification of patterns, but his analysis, if it is to be more than descriptive, must proceed from there to the study of the interrelationships which are responsible for those patterns. It is at this point that he moves more directly into those systematic fields from which his original data were taken. Even locational analysis, which would appear to be most concerned with pattern for its own sake, is subject to such a view. Haggett's attempt to isolate a separate set of interests comprising the geometrical aspects of geography is thus mistaken in so far as the real interest in the existence of a pattern in, say, the distribution of central places, is not the existence of the pattern itself, but the understanding of the movements and circulations which are responsible for the spatial pattern.[3] To the social geographer working in urban areas, spatial patterns are thus what they are for the ecologist—a reflection of social processes which are at once highlighted and better understood by the identification of the spatial distributions and the spatial associations.

The second recurring theme is that of the process of adaptation. For the ecologist this is a process of multiple interaction between man and environment and between man and man. Again, the geographical approach can be seen to be identical. To a large extent it is an accident of history that it has been the man/environment complex which has been stressed in geography. For long, geography was concerned with the identification of the natural or formal region and the increasingly rural orientation which this was given by the despairing search for the perfect *pays* served only to emphasize the false dichotomy between the human and physical aspects and the concentration on their interaction, whether seen in terms of environmental effects on man in the determinists and possibilists, or in terms of man's effect on environment.

[1] Duncan, 'Human ecology and population studies', p. 681.
[2] Ackerman, *Ann. Ass. Am. Geogr.* LIII (1963), 437.
[3] P. Haggett, *Locational analysis in human geography* (London, 1965), pp. 13-16.

The fact that ecology was developed within the realm of sociology and that, in contrast to geography, its most productive advocates concentrated their efforts on urban areas, helped to shift the focus of interest within ecology from physical environment/human reactions, to one which emphasized man's relations with man. Today, geographers no longer consider environment to be the nebulous concept of 'natural environment'. Hartshorne, for example, has shown how it is conceptually impossible to separate a so-called natural environment from a human-influenced environment. Increasingly a position has been accepted which is the only logically tenable one: namely that environment is everything that is external to the environed unit. This is what the Sprouts have termed the 'milieu'.[1] As Ackerman, Trewartha and others have noted, cultural phenomena and other human individuals are as much a part of an individual's environment as the nature of the physical phenomena which form the backdrop. For the individual, or for the human group living in a spatially circumscribed territory, the fact of high or low population density is as important as, if not more important than, the nature of the soil or the absence or presence of mineral resources within their area. The very term 'environment' presupposes the existence of something which is environed and, for that environed unit, whether it is an individual or an aggregate, the 'environment' must comprise everything that is external to it. Physical and non-physical, organic and inorganic, tangible and intangible, all are included if external to the environed unit. It is upon this concept that most of the work of population geography and certainly most of the methodological pleas for population geography have been based.[2]

Geography, too, must therefore be concerned, and indeed is concerned, with the relations of man to man and human group to human group. As Hartshorne concludes in his later writings on the methodology of geography, 'Geography, in seeking to analyze the complexity of integrated phenomena in reality, is concerned to examine relationships among phenomena, of whatever kinds, which are found to be significant in the total integration. In many cases, such relationships may be those between human and non-human phenomena, in others between animate (whether human or non-human) and inanimate phenomena, or between visible and invisible, or between material and non-material. But no one of these dichotomies is logically of any more significance to geography than any other.'[3] The logical extension is that the study of the interrelationships of sets of phenomena within any one of these

---

[1] H. and M. Sprout, *The ecological perspective on human affairs: with special reference to international politics* (Princeton, 1965).

[2] In this connection, the impressive expansion in the number of works falling within population geography illustrates the increasing concern of geographers with human phenomena. See, for example, W. Zelinsky, *A bibliographic guide to population geography*, University of Chicago, Department of Geography Research Paper no. 80 (Chicago, 1962).

[3] Hartshorne, *Perspective*, pp. 63–4.

categories also falls within the scope of geographical analysis. Those writings which attempt to differentiate ecology and geography in terms of the latter's emphasis upon environment and the former's emphasis upon man, thus adopt a mistaken view of the nature of geography. In essence, all that has been said and all that has been examined has suggested the very close similarity between geography and human ecology in their approaches to the study of man, and particularly to urban analysis.

THE ROLE OF CULTURE

Within such an ecological geography there still remains, however, the question of the role of cultural factors, as against material factors, in the explanation of human organization and behaviour. That a pattern exists in urban structure cannot be denied. That this pattern is independent of culture is, however, a dubious proposition. One of the strongest criticisms of the empirical work of the Chicago ecologists is that it has proved time-bound and culture-bound: that what may hold true for the American city at one particular time cannot be generalized to form the basis of prediction or analysis for cities at different periods and in different culture areas. The most striking examples of studies which have illustrated this have been drawn from Latin America where, for example, the Burgess pattern has been shown to be reversed, with high status areas being found towards the centre of cities and low status areas at the periphery. Sjoberg, from a wide-ranging review, has suggested that such divergencies are related not so much to regional effects as to a universal distinction between pre-industrial and industrial cities and that the 'American' pattern is beginning to be superimposed wherever the values and the structure of industrial civilization are being overlaid upon older foundations.[1] The important point is that the nature of a society's values affects the form of its organization and finds spatial expression in such things as the structure of its towns. The analyses of the Chicago ecologists might therefore be viewed as a particular set of 'laws' which are applicable only within the context of the unstated assumptions underlying their approach to the city.

There is a large number of such assumptions. First, there are certain contextual variables which have to be borne in mind. The fact that the cities at the time, and especially Chicago, were being continuously deluged by immigrants of a variety of ethnic types provided a population matrix which was important in the derivation of ecological theories. These waves of immigrants were particularly heavy in the period 1890 to 1910. There was

[1] G. Sjoberg, 'On the spatial structure of cities in the two Americas', in Hauser and Schnore (eds.), *The study of urbanization*, pp. 347–98.

also the vitally important fact of the startling growth of Chicago. In 1860 its population had been 110,000; by 1870 it had risen to 300,000. It doubled from 500,000 to over one million between 1880 and 1890 and by 1910 it stood at over two millions. The Burgess model must be viewed as an attempt to deal with the organization of spatial phenomena which occurs within the context of growth and expansion. The very title of his original essay—'The growth of the city'—shows his own awareness of the importance of expansion and growth. Quinn has also pointed to some other assumptions underlying the Burgess hypothesis.[1] These include the existence of heterogeneity within the city, the existence of a particular kind of economic base (that of mixed commercial-industrial city), and the existence of certain economic and cultural factors such as the private ownership of property and an efficient transport system equally rapid and cheap in every direction within the city. In addition to these contextual variables and assumptions, whether overt or unacknowledged, there are, secondly, some important value orientations which are implicit in the scheme. These include the following: an assumption of the motive of profit maximization and hence the value which business enterprises place on central location because of its short supply and greater accessibility to the whole of the urban area; the social value which is placed on residential location at the periphery of the city which, presumably, derives from the value which is accorded to access to open space and the possibility which such location offers for the building of large houses with large gardens: and also such institutional factors as the existence of economic competition, relatively free from institutional control. These are but some of the assumptions which relate to values. The fact that they were assumed as constants in the Burgess scheme is a feature which can rightly be criticized. If it were possible to incorporate such elements within a model of urban structure, the Burgess scheme could be re-expressed in terms of a probabilistic rather than a deterministic model and so incorporate those distortions which gave rise to the criticisms of such scholars as Firey.

Analysis of urban structure must therefore recognize the value assumptions upon which it is based and, if possible, incorporate such aspects as social and psychological influences on those decision-making processes which lead to the development of spatial pattern in towns. As Ackerman has noted, 'Influenced by anthropological thought, many cultural geographers are coming to realize that ideas, attitudes, and other non-visible entities of a culture are of importance in understanding spatial distributions and space relations of phenomena'.[2] The importance of attitudes and value orientations

[1] J. A. Quinn, 'The Burgess zonal hypothesis and the critics', *Am. Sociol. Rev.* v (1940), 210–18.
[2] E. A. Ackerman, *Geography as a fundamental research discipline*, University of Chicago, Department of Geography Research Paper no. 53 (Chicago, 1958), p. 28.

is that they act as intermediaries within the relationship between environment and environed unit. The beliefs and values which the individual has internalized or which the group of individuals holds, alter the individual's or the group's evaluation of environmental stimuli and environmental potential.[1] In order to understand the relationship between environment and environed unit it is therefore essential to be able to take account of these values and attitudes. This was the weakness of Barrows' proposals. Since they ignored the consideration of the broad sociological aspects of the spatial interrelationships of human activities they were incapable of developing powerful concepts. In the 1920s sociology had not developed a sufficiently coherent corpus of theory to enable such factors to be taken into account, but the situation is radically different today.

The Sprouts, for example, have differentiated between what they call a 'psycho-milieu' and an 'operational milieu'; in other words between 'the milieu as perceived and reacted to by a specified individual, and the milieu in its total relations to that individual as this would appear to an observer who sees all and knows all'.[2] They suggest that any model of decision making has to incorporate within it certain basic assumptions relating to three aspects of the individual or group: first regarding motives, second regarding the knowledge of the operational milieu, and third regarding the competence to carry out the stated objectives. A 'common-sense' model assumes a profit motive, adequate knowledge of the operational milieu and rationality in choosing ends and means to achieve the objectives. It has been shown how certain of these assumptions, albeit unstated, characterized the Burgess scheme. What is needed in analysing urban structure is a conscious attempt to include differing types of value orientations as intervening variables. The individual, who may or may not form a unit within a social group, perceives his total external environment in terms of the attitudes and values which he has internalized and which may derive in large part from the group of which he may be a member. Analysis and prediction of social structure will become increasingly sharpened as it becomes possible to incorporate more facets of these cultural components within probabilistic models. It is for this reason that there is value in trying to weld elements of sociology into our ecological perspective on social geography. The substantive work which is discussed later illustrates the approach of such a composite analysis, which draws within its compass the analysis of spatial distributions and the study of attitudes. To anticipate this, a concrete example of the role of attitudes might be suggested. Sociologists suggest a number of modes of value orientation

[1] The geographer Kirk has advanced a similar argument in showing the relevance of *Gestalt* psychological theory for geographic analysis. Cf. W. Kirk, quoted in H. and M. Sprout, *An ecological perspective, op. cit.* pp. 130–1.
[2] H. and M. Sprout, *op. cit.* p. 136.

amongst which two might be noted: the non-traditional and the traditional. The non-traditional is purposive and rational. The individual makes decisions in terms of a set of explicit criteria which lead towards some goal, such as power, prestige, profit or security. Typically, such a set of attitudes and behaviour might be found within what might roughly be designated as the middle class, or perhaps, the aspirant elements within the working class. It is just such an orientation on which most 'common-sense' models of behaviour are based and includes some of the unstated assumptions included within the framework of the classical ecologists. The traditional orientation, by comparison, is less individualistic, more dependent on the authority of a leader or a venerated book or, what is more significant here, upon the authority of group values and group constraints. Much of what Firey called sentiment and symbolism is included within this concept of traditional orientation, whether it be the 'illogical' adherence to the traditions of Beacon Hill or the maintenance of Italian group traditions in the North End. Much of the behaviour which might be labelled irrational or illogical is simply the outcome of frames of reference which differ from the rational mode which is the usual construct around which most models are built. Stacey, for example, uses this dichotomy to study social structure within Banbury and demonstrates its importance in understanding such socio-economic characteristics as voting behaviour, types of occupation, religious affiliation, family relationships and so forth.[1]

As Beshers has shown in a stimulating study of urban social structure, the viewpoints of human ecology and of functional sociology can be blended to produce a methodological framework for an overall view of the organization of urban social systems.[2] Spatial distributions can be viewed as a mechanism which at once reflects and perpetuates social structure. It is the later purpose of the present study to show how such interconnections between spatial distributions, social structure and social attitudes can be exploited to aid understanding of urban life. 'It might well be maintained', as Caplow said, 'that the need for a close liaison between sociology and geography in the study of city life existed from the beginning, and that the curious isolation of the two literatures from each other is an unfortunate accident.'[3] The examination of the fields of geography and human ecology has provided a framework within which an empirical examination of the nature of the relations between space and urban social structure can be attempted.

[1] M. Stacey, *Tradition and change: a study of Banbury* (London, 1960).
[2] J. M. Beshers, *Urban social structure* (New York, 1962).
[3] T. Caplow, 'Urban structure in France', *Am. Sociol. Rev.* XVII (1952), 549.

CHAPTER 2

URBAN SOCIAL STRUCTURE:
DATA, TECHNIQUES AND ANALYSIS

Data and techniques must be at the very root of any attempt to analyse the nature of the spatial structuring of urban areas. Given the size and given the intricate system of interdependent elements which the city represents, it is obvious that the type, the detail and the accuracy of the material which is used to describe the city largely condition the ideas and the theories which must emerge from any empirical approach at analysis. Further, the techniques which are used have a direct bearing on the theories which are likely to emerge. Technique can never be thought of as a substitute for theory, but at an empirical level, certain types of technique are likely to lead to more potent theoretical statements than are others.

It is for this reason that this chapter examines some aspects of census data and of the areal units for which such material is presented. Before urban analysis can develop reliable theory it is essential that interested urban scholars clarify their needs for census data and make them known to the census authorities. Having considered the basic data, the chapter then examines certain alternative techniques in the spatial analysis of urban society. While the technique of Social Area Analysis appears to have grave limitations, one of the most promising of recent techniques is multivariate analysis, and the results of the few urban studies which have used this method are compared. Finally, some of the problems involved in the aggregation of areal units into urban social regions are discussed, since this is one of the end products of the application of multivariate analysis.

THE CENSUS

The complexity of urban social phenomena forces the student of the city into a dangerous reliance on census data. There is irony in this since complexity ought to call forth more sophisticated data-collecting geared to the specific needs of the investigation in hand, but the sheer quantity and variety of material which is needed rule out the use of direct field observation and data collection for any but the most limited survey of urban analysis. It is significant that the early years of the Chicago school's work were dominated by the need to collect data and that one of the very early projects which was undertaken was the production of the *Local community handbook of Chicago* which first appeared in 1930 and has been re-issued after each subsequent census, up to 1960. Census data is collected for a variety of purposes. Certainly it is not prepared with the main object of meeting the needs of urban research. In Britain, from the early concentration on population data alone, the bias has swung increasingly to meeting the needs of city planners as is suggested by the great wealth of detail on the physical condition of houses.

The Census Office, however, has recently begun to recognize the needs of urban research. From a concern with the production of census volumes as ends in themselves, the tendency now seems to be to give greater attention to the uses to which the data can be put. The greater ease of access to detailed information and the consultation with such bodies as the Institute of British Geographers or the Inter-University Census Tract Committee in seeking suggestions as to boundary problems and as to which questions should be included in census schedules, are manifestations of this change of attitude, which are both welcome and long-sighted.[1] The lack of material suitable for research purposes may well have been due to the fact that academics have not made their needs clearer, but the change is certainly much needed. Particularly is this so in the case of the areas within which census statistics are presented. The three basic areal units which are subsequently discussed are as follows: the *ward* which is an administrative unit; the *enumeration district* (e.d.) which is the small area of households for which census data are collected by one enumerator; and the *census tract* which is formed of a combination of e.d.s with the aim of ensuring that the population in each tract is as homogeneous as possible. In Britain, population data for wards have been available over a considerable period of time. The first example of comprehensive small-area data, however, was in the tracts devised for Oxford for the 1951 census.[2] With the 1961 census, the use of tracts was abandoned in favour of the decision to release enumeration district material wherever it was specially commissioned.[3]

For long, therefore, the smallest areal unit which was used for the publication of data was the ward, and even here it was usually only population or

[1] The Inter-University Census Tract Committee—originally established to consider the delimitation of Census tracts—has altered its focus to act now as a multi-disciplinary body drawing together scholars concerned with the analysis of urban census data and the study of urban social structure. Its secretary is Mr Peter Norman, Department of Social and Economic Research, Glasgow University.

[2] Oxford Census Tract Committee, *Census 1951, Oxford area: selected population and housing characteristics by census tracts* (Oxford, 1957).

[3] See General Register Office, *Census 1961, statistics for wards, civil parishes and other small areas.* Circular C 61/12 (July 1961).

40

density data that were available. Such areas are not only too large to provide the necessary web of detail for urban research, but the wards themselves, being administrative units, are only too often meaningless aggregates in terms of the social and physical characteristics which they contain.

The decision by the Registrar General to release enumeration district data in 1961 was partly due to the proselytizing work of Collison and others in Oxford.[1] For the first time a wealth of detailed information was released which has already proved fruitful in widening the scope of urban research. A few workers had succeeded in obtaining enumeration district data before 1961, but the cost and difficulty of doing so had deterred many from following their example.[2] Today, with the possibility of buying enumeration district data from the census office, there is a great need for data banks of the material to be built up for a variety of types of towns. Local authorities seem best placed to meet this need and many local authorities did indeed obtain such data from the 1961 census, but since such collection of small area data depended entirely on the interest and progressiveness of the local authorities concerned the coverage is haphazard. For future censuses, there is a case for encouraging and perhaps assisting a more systematic and complete collection of data by local authorities. Not only would this ensure ready access to data for research purposes in future years, but would provide planning departments with data of sufficient detail and variety for a multitude of purposes, since the enumeration districts, each of which includes an average of well below 1,000 population, provide a fine enough areal mesh to supply invaluable local information for planning purposes.

The situation in the United States has been somewhat different from that in Britain. The interest which has long been shown in urban areas by American scholars has undoubtedly contributed to the much more intimate association of academics and census officials within America. The fact that such scholars as Schmid have assisted in the work of the Census Bureau, that it includes a Geographic Division and that numerous research projects have been carried out under its auspices all illustrate the healthy interchange between research studies and the production of census data. This is particularly the case as regards the areal presentation of statistics. Unlike Britain, small-area data have been available for a considerable time in America, in the form of census tract and block statistics. American census tracts are small, permanently established geographic areas into which large

---

[1] Oxford Census Tract Committee.
[2] Prior to 1961, the only cases of the use of e.d. known to the author are: R. Glass, *The social background of a plan: a study of Middlesbrough* (London, 1958); J. M. Mogey, *Family and neighbourhood: two studies in Oxford* (London, 1956); P. Collison and J. Mogey, 'Residence and social class in Oxford', *Am. J. Sociol.* LXIV (1959), 599–605; P. Collison, 'Occupation, education and housing in an English city', *Am. J. Sociol.* LXV (1960), 588–97; E. Jones, *A social geography of Belfast* (London, 1960). In each case, data were specially extracted by the General Register Office.

cities and their environs have been divided for statistical purposes. Their boundaries are established by local committees and approved by the Census Bureau. The average tract contains some 4,000 people and is originally laid out 'with attention to achieving some uniformity of population characteristics, economic status and living conditions'.[1] They were first established for the 1910 census for the cities of New York, Baltimore, Boston, Chicago, Cleveland, Philadelphia, Pittsburgh and St Louis, all of which were tracted. Subsequent censuses have included growing numbers of tracted cities: ten more were added for 1930, 42 more for 1940 and the 1950 census recognized tracts for 115 cities together with adjacent areas around some of the cities. For the 1960 census, there were no fewer than 180 tracted areas, of which 136 were entire Standard Metropolitan Statistical Areas.

The delineation of these census tracts has been the responsibility of local committees which may be formed from local bodies such as the chamber of commerce, the council for social agencies or an advisory committee of the planning commission, or from such professional bodies as the American Statistical Association, the American Marketing Association or the American Institute of Planners. The local committee submits suggestions to the Census Bureau on the basis of certain criteria. For example, given a city of over 50,000 population or a Standard Metropolitan Area with a central city of over 100,000, the size of tracts can vary between 2,500 and 8,000 population, except for the Central Business District tract or possibly some institutional tracts which may be smaller. The boundaries ideally follow permanent recognizable lines and contain, where practicable, people of similar racial and economic status and areas of similar housing. The problem and the principles on which such boundaries can be drawn have in fact given rise to a considerable body of literature in the United States.[2]

On the basis of such tracts, a number of functional groupings have also been made. The Central Business District (C.B.D.) is one such area which was first designed for use in the 1954 Census of Business when 94 cities of over 100,000 population had C.B.D.s designated either as a tract or group of tracts. Statistics showing types of business establishments, sales and payrolls in both 1948 and 1954 were published for these C.B.D. areas. A second, related, grouping of tracts is the 'retail trade area'. The C.B.D. is one such retail trade area, but others lying outside the C.B.D. have been delineated to include areas containing 400 or more stores. A third functional grouping is the 'census county division' which was first established by Schmid in

---

[1] *Census tract manual* (Bureau of the Census, Washington, D.C., 4th edition, 1958), p. 1.
[2] Some methods are discussed for example in W. H. Form *et al.*, 'The comparability of alternative approaches to the delimitation of urban sub-areas', *Am. Sociol. Rev.* XIX (1954), 434–40. See also C. F. Schmid, 'The theory and practice of planning census tracts', *Sociol. Soc. Res.* XXII (1938), 228–38.

Washington and has since been extended to a number of counties. Whereas tracts are formed of units of similar economic and social character, the C.B.D. forms a unit which is a community or group of communities. Each usually has a central place as its focus.

These are some examples of the functional, as against administrative, areas within which statistics are prepared and published for relatively small areas in the United States. In addition to the tracts, there are also data in certain cities for blocks of houses. Block statistics from the Census of Housing were first published in 1940 for cities of over 50,000 population, and for the 1960 census certain cities of below this size approached the Census Bureau for block data. Statistics published for each block include the classification by dwellings of type of occupancy, the condition of plumbing facilities, persons per room, the colour of occupants, and the monthly rent or average value for the rented and owner-occupied sectors respectively. For tracts the range of published data includes, from the Population Census, population, age, sex, race, marital status, education, residential mobility, income and occupation, and from the Housing Census, tenure, type of structure, age, density of population, presence or absence of such facilities as television or refrigerator, type of fuel used, and rent or value.

Small-area data and data presented for more meaningful areas than administrative units have therefore been much more readily available within America than in Britain. Some idea of the widespread interest which has been shown in this material can be gauged from the fact that, while the Census Bureau only publishes tract data for those areas which show positive interest in its use, tract data in the 1950 census were not published for only six of the total tracted areas.[1] Demand for the block statistics is shown by the fact that 200 communities of under 50,000 (the lower ceiling limit for getting free data) paid the cost of special data tabulation for 1960.[2] Foley has illustrated the wide use to which small-area data have been put, both in pure and applied research. In the field of pure research, however, he suggests that only the surface potential of tracts had been tapped up till then, mainly by sociologists. Of geographical work, he comments, 'Urban geographers have relied on their own mapping and have generally shunned the comparative, quantitative methodology that would most logically provide a receptive context for using census tract data'.[3] Nevertheless, the list of pure research which has been made possible by the development of tract statistics makes impressive reading. For example, Stouffer's important methodological work

[1] *The 1950 censuses : how they were taken*, Procedural studies of the 1950 censuses, no. 2 (Bureau of the Census, Washington, D.C., 1955).
[2] F. S. Kristof, 'The increased utility of the 1960 housing census for planning', *J. Am. Inst. Planners*, xxix (1963), 40–7.
[3] D. L. Foley, 'Census tracts and urban research', *J. Am. statist. Ass.* xlviii (1953), 734.

on migration and intervening opportunities was only made feasible by the existence of prior work by Green on inter-tract population mobility, using tract data for the city of Cleveland.[1]

This American experience illustrates the value and importance of small-area census data and the recent change of attitude on the part of the British census is therefore both welcome and overdue. Workers in both pure and applied urban research should be encouraged to use their influence to further the collection and release of small-area data and of material which is more suited to research needs. Certainly, there is a great need for more informed criticism of census data. Comparison with the American case, for example, suggests that there is still a great deal of scope in Britain for improving the type of data and the areal presentation of census information. One realm in which this is particularly true is in the need to make greater use of functional rather than administrative areas wherever this is feasible.

A question which arises in the British context is whether census data are more valuable in the form of enumeration districts or for the larger census tracts. Arguments might be advanced for either case. In Oxford, where the prototype of British census tracts was developed, it was suggested that tracts should not be greatly above 3,000 in population and in fact the 48 tracts which were delineated for Oxford had an average size of 2,645.[2] While smaller than the American version, these are nevertheless fairly large aggregates, and likely in consequence to include a high degree of heterogeneity irrespective of the criteria used in their delimitation. In America, for example, Myers has noted that out of 28 tracts in New Haven, 10 were remarkably homogeneous, 7 were less homogeneous and 11 were heterogeneous.[3] While this is an obvious and unavoidable consequence of using a basic unit of some 4,000 (in the American case) or 3,000 (in the British case), the greatest advantage of the tract which has been put forward is its stability over time. By adopting large units, it is argued, it is possible to maintain areal boundaries and so permit comparisons over time. This is perhaps an optimistic hope. With constant internal change, particularly in the present period of central redevelopment, the maintenance of stable tract boundaries might be possible but unwise. Areal units which were originally delineated as homogeneous may be unlikely to stay so for very long. Perhaps a more valid argument in favour of tracts is the possibility of developing the collection or recording of non-census data within the tract boundaries. In America, Foley notes that local bodies and researchers have collected statistics relating

[1] S. A. Stouffer, 'Intervening opportunities: a theory relating mobility and distance', *Am. Sociol. Rev.* xv (1940), 845–67: H. W. Green, *Movements of families within the Cleveland metropolitan district* (Cleveland, Ohio, 1936).
[2] Oxford Census Tract Committee, *Census 1951, Oxford Area.*
[3] J. K. Myers, 'Note on the homogeneity of census tracts', *Social Forces,* XXXII (1954), 364–6.

to juvenile delinquency, receipt of welfare care, births and deaths, illness, mental illness, suicide, and mobility within the framework of the census tracts.[1] In Oxford, an attempt was made to present data on rateable values for the tracts. Given tracts of relatively large population, it is possible that local authorities might assemble data on similar topics and on such rates as crime. It is much less likely that they would consider this for smaller areas such as enumeration districts which would, in any case, be much more subject to changing inter-censal boundaries.

On the whole, however, the Registrar General's decision to make enumeration district data available instead of developing a widespread use of tracts is to be welcomed. The greater detail which they reveal and the greater likelihood of demarcating homogeneous areas, which they inevitably present, are to be preferred to the dubious virtue of stability over time of the tract. Since the enumeration districts are so much smaller, it is always possible to aggregate individual enumeration districts so as to preserve a rough stability as between census periods. As to the collection of other types of data within their boundaries, the fact has to be faced that any widespread or systematic collection of such data by local authorities is slight in the extreme and the individual researcher will do better to rely on his own aggregation of such data. The finer areal mesh which the enumeration district offers gives it a flexibility and utility which overrides the advantages offered on behalf of the census tract.

In passing it might be noted that considerable advance appears possible in the case of the 1971 British census. Not only are outside bodies to be consulted, but the Census Office is establishing advisory panels to consider such aspects as demography, urban studies, housing and regional studies. On the vital question of areal units, it appears that two important innovations may be introduced: first, that data may be aggregated for meaningful areas (such as built-up zones rather than administrative units) in the form of 'locality statistics'; second, that data may be presented for 100-metre grid squares on the National Grid. (Information from Mr. F. E. Whitehead, statistician of the General Register Office, at a meeting of the Inter-University Census Tract Committee, June 1967.) The grid-square principle would solve many of the technical problems confronting the use of areal data (see, for example, pp. 157–9 below) and would be a most exciting and welcome addition to the enumeration district. Grid-square statistics have, for example, helped greatly in the seminal work done in Sweden.[2]

---

[1] Foley, *J. Am. statist. Ass.* (1953).
[2] See, for example, T. Hägerstrand, 'The computer and the geographer', *Transactions, Institute of British Geographers*, 42 (1967), 1–19. On prospects for the British 1971 census, see 'Grid Squares for Planning: possibilities and prospects' papers of the Annual Conference, Regional Studies Association, October 1967.

Urban analysis

MULTIVARIATE NATURE OF THE CITY

The variety of data which is provided by census enumerations has widened enormously with time. Each new census has tended to introduce a new set of statistics providing more precise parameters for topics previously not covered or inadequately covered. Today the census gives a broad cover of data relating to demographic, social, economic and housing characteristics. The inclusion of questions on possession of motor cars and garaging and on the type of transport involved in the journey to work, for example, were a new feature of the 1966 sample census. This proliferation of topics does some justice to the organized complexity of social characteristics. The multivariate nature of urban areas cannot be subsumed under any single parameter. Rather, in looking at a given individual or aggregate of individuals, the analyst is interested not only in numbers, but also in age, sex, occupation and a host of other features. The wealth of data therefore allows the necessary breadth of information to be considered. It also presents problems of interpretation and handling. The difficulty is one of trying to include as much detail as possible and yet to handle this detail as parsimoniously as techniques permit and as theory suggests valid. The correlation matrix is the usual point at which most statistical observations of social structure begin. The matrix of correlation coefficients shows how each characteristic or variable is related to each other in turn, but an examination of most matrices shows that many of the variables tend to duplicate other variables; that the picture which is revealed by, say, the proportion of white collar workers in an area is virtually the same as that shown by, say, the proportion of persons with a good deal of formal education. Price provides a suggestive illustration of this duplication. His matrix of correlation coefficients shows a coefficient of 0·47 between the sex ratio and the median rental of areas. Neither common sense nor theory would suggest that there is a causal association between the two. Both in fact correlate highly with his measure of wages and he shows that a partial correlation between the sex ratio and rents, holding wages constant, is almost zero. Therefore the correlation between the sex ratio and rental levels can be suggested as being 'explained' in terms of their association with wages.[1] In this system or universe of three variables the level of wages can therefore be thought of as the fundamental variable. Given a much larger universe of variables describing aspects of social structure, analysis of the salient features of that structure can best be accomplished by seeking out systems of such fundamental variables which will explain the patterns of interrelations which are often concealed by the individual correlations.

[1] D. O. Price, 'Factor analysis in the study of metropolitan centres', *Social Forces*, XX (1942), 449–55.

46

It is along such lines that analysis of social structure through the census material has proceeded. An interest in the geographical or spatial elements of urban social structure inevitably leads to the attempt to isolate sub-areas within towns which can reveal the internal structuring of the urban population. The contribution of the early ecologists to this field has already been examined. In this section the more recent techniques and approaches to the formation of urban sub-areas will be discussed.

SOCIAL AREA ANALYSIS

The differing lines of approach have largely been determined on the basis of varying theories of social structure. One of the most influential approaches has been that of Social Area Analysis[1] as developed by Eshref Shevky and his associates.[2] Shevky's technique suggests that the complexity of urban social structure can be subsumed under three basic constructs which are termed *social rank*, *urbanization* and *segregation*. Originally, these three indexes were composed of seven individual variables as follows:

$$social\ rank \begin{cases} \text{occupation} \\ \text{education} \\ \text{rental} \end{cases} \quad urbanization \begin{cases} \text{fertility} \\ \text{women in labour force} \\ \text{single-family dwellings} \end{cases} \quad segregation$$

Social rank is a compound of occupation, education and rental levels with high rank being indicated by high proportions of non-manual workers, high proportions of persons with long formal education and high rental levels. *Urbanization* is composed of fertility, women in the labour force and single-family dwellings. *Segregation* is a measure of the concentration of ethnic or minority groups in a given area in relation to their proportion in the urban area as a whole. By standardizing and combining the scores of each of the sets of three variables which form the first two indexes, sub-areas in towns can be allotted to a position within a two–dimensional 'social space' diagram whose axes are *social rank* and *urbanization*. These sub-areas can then be differentiated further in a third dimension according to whether they have high or low *segregation* scores. Shevky's typology thus isolates and classifies urban sub-areas in terms of the three basic constructs which are believed to be necessary and sufficient parameters of urban social structure.

Its authors' suggestions for the use of this typology indicate the breadth of

[1] 'Social Area Analysis' is a technical term used to describe the Shevky technique. Since confusion might arise between the use of the term in this technical and restricted sense on one hand and its use to describe the generic study of social areas on the other, the words 'Shevky's technique' or 'Shevky's typology' are used in place of 'Social Area Analysis'.

[2] The two principal statements are E. Shevky and M. Williams, *The social areas of Los Angeles: analysis and typology* (Berkeley, 1949), and E. Shevky and W. Bell, *Social area analysis: theory, illustrative application and computational procedure* (Stanford, 1955).

scope which is claimed for the method. One suggestion, for example, is that the technique can be used to analyse units other than urban census tracts: whole cities could be used as units or, alternatively, counties or even whole countries should be just as amenable to the technique. The scope of the concept is therefore not restricted to the urban area. Rather, the three constructs are claimed effectively to summarize the total society as well as its cities. Other suggested uses include the comparative study of different areas at one point in time, in which Shevky's technique could be used to determine 'particular recurrent space-time-value patterns' which, it is hypothesized, would occur despite variations in the pattern of internal social differentiation of the areas.

The delineation of urban sub-areas and the use of such areas as a framework for further research are the two other uses which have been pursued most fully in the published literature and which are most pertinent to our spatial interests here. Using the technique, urban social areas are formed by aggregating units on the basis of their similarity with respect to Shevky's three basic indexes. In essence such social areas are rather similar to the 'natural areas' of the classical ecologists since they are seen as containing 'persons with similar social positions in the larger society'. There are differences between the two concepts however: 'The social area...is not bounded by the geographical frame of reference as is the natural area, nor by implications concerning the degree of interaction between persons in the local community as is the subculture. We do claim, however, that the social area generally contains persons having the same level of living, the same way of life and the same ethnic background, and we hypothesize that persons living in a particular type of social area would systematically differ with respect to characteristic attitudes and behaviour from persons living in another type of social area.'[1]

Since its introduction, there have been a number of changes made to the technique and a number of disagreements even between its proponents. In later studies, for example, the measure of rental is dropped from the *social rank* index. The social space diagram too is changed: from one with nine 'social spaces' based on the regression line of *urbanization* on *social rank*, to one with sixteen social spaces formed by quartile divisions of both axes. A third disagreement is Bell's substitution of the terms 'economic status', 'family status' and 'ethnic status' in place of Shevky's original nomenclature for the three indexes. This would appear to be a reflection of Bell's disagreement as to the level of generality at which the Shevky analysis can be applied.

The method indeed has become a source of great controversy wherever it

[1] Shevky and Bell, *Social area analysis*, p. 20.

has been used. Following the original formulations, a number of studies have examined the validity and generality of the constructs. Obviously, one of the most basic questions is whether or not the two most basic indexes of *social rank* and *urbanization* are unrelated as Shevky suggests and whether they do in fact measure different aspects of social organization. The concept of *social rank* is relatively familiar in the literature of sociology since it is closely parallel to the concept of socio-economic status and is composed of phenomena which have long been regarded as being related to social status. The concept of *urbanization*, however, is less familiar. Essentially it would seem to owe a good deal to the ideas of Louis Wirth.[1] Shevky and Bell consider that 'Wirth's ideas were too narrowly conceived and should be related to the analysis of a total society as well as of its cities'.[2] Nevertheless, the concept of *urbanization* is evidently closely related to Wirth's idea of the decreasing importance of primary contacts within the city and the decreasing role of the family as a social unit. To this extent, *urbanization* is somewhat akin to Durkheim's 'organic solidarity', Redfield's concept of 'urban' traits, or Tönnies' *gesellschaft*. The question is whether such concepts developed at a societal level can be translated to meaningful differences within an urban area and whether Shevky's index of *urbanization* is an effective measure of such assumed differences.

On the question of the validity of the two indexes, a number of West Coast sociologists have attempted to justify the empirical separateness of *social rank* and *urbanization*. Bell himself has used factor analysis as a means of testing two hypotheses: first, that economic, family and ethnic status each represent discrete factors which are necessary to account for the social differentiation of urban sub-populations; and secondly that the two sets of data used to compose 'economic status' and 'family status' are each uni-dimensional indexes.[3] His findings lend support to the hypotheses and represent a partial validation of the Shevky technique. A more interesting, although not independent, study which provides substantive support is the application of cluster analysis by Tryon.[4] Cluster analysis, as developed by Tryon, is a statistical procedure which delineates the general properties of a group of objects and then groups the objects into homogeneous types in terms of these properties. In very simplified form, it is analogous to factor analysis. Tryon's study of San Francisco represents an attempt to derive social dimensions empirically and to classify people—or areas—into sub-

[1] L. Wirth, 'Urbanism as a way of life', *Am. J. Sociol.* XLIV (1938), 1–24.
[2] Shevky and Bell, *op. cit.* p. 8.
[3] W. Bell, 'Economic, family and ethnic status', *Am. Sociol. Rev.* XX (1955), 45–52.
[4] R. C. Tryon, *Identification of social areas by cluster analysis* (Berkeley, 1955). The fact that this study and the work of Shevky are not independent is suggested by Tryon's preface (p. iii) where the cross-fertilization of ideas between Shevky, Bell, Schmid and Tryon is noted.

cultural groups of types in terms of these dimensions. It thus avoids some of the dangers of choosing dimensions on *a priori* grounds. From his initial 33 variables, which measure population characteristics, occupational characteristics and housing characteristics, Tryon concludes that 29 of the variables can be grouped to form 7 clusters of related measures which are 'capable of reproducing virtually all the intercorrelations among the 33 measures'. Examination of the intercorrelations between these 7 clusters suggests that three independent dimensions are necessary and sufficient to account for the intercorrelations and the three most independent which are selected are labelled as follows: first, 'family life', which characterizes tracts with well-developed family existence so that high scores indicate more housewives, more owner-occupied single-family homes, few women at work and larger families; second, 'socio-economic independence', which is a dimension of wealth and social independence and has high scores where tracts have high proportions of persons who are self-employed, have servants, expensive homes and a college education; and third, 'assimilation', which measures the incorporation of persons into the 'standard white-collar American culture' and thus has high scores where tracts have few people who are foreign-born, especially those from underdeveloped countries, few non-whites, a high proportion of females and a small proportion in manual occupations.

Tryon notes that there is considerable similarity between these empirically derived clusters and the three indexes which Shevky isolates. *Social rank*, as conceived by Shevky, is similar to a cluster which Tryon calls 'socio-economic achievement' which, while not considered necessary to the general description of urban social structure, is closely related to 'socio-economic independence'. *Urbanization* is closely parallel to Tryon's cluster of 'family life'. *Segregation* includes data which are incorporated into Tryon's 'assimilation' cluster, but is a much more narrowly defined concept, and he suggests that the validity of the measure would have been increased by including a greater number of measures. 'Assimilation' is thus preferred to *segregation* both on conceptual and measurement grounds. Noting the partial similarity between the two sets of results Tryon comments: 'Shevky and Williams showed considerable insight in thus choosing on purely *a priori* grounds three dimensions partly resembling three of the four we derived empirically...However, social rank, urbanization, and segregation do not adequately cover the ground. The more independent dimension, socio-economic independence, is missing from their theory.'[1] Furthermore, he suggests that the validity of each of their measures could have been increased by the inclusion of more variables.

Tryon thus adds a certain degree of empirical justification to the approach

[1] Tryon, *Identification of social areas by cluster analysis*, p. 13.

of the Shevky analysis. His method of cluster analysis, however, leaves a great deal to be desired. Its basic similarity to factor analysis has been noted. Tryon's justification for preferring cluster analysis is that it is a much simpler method of isolating sets of interrelated phenomena. With the relative ease of access to computers which scholars now have, this is now more of a hindrance than a help. For example, in determining the composition of his clusters, Tryon uses a graphical method of comparing the pattern of intercorrelations amongst the sets of variables. This has the advantage of simplicity, but is more subject to individual interpretation in the making of any selection. It also leaves the weights of each variable within the cluster—what Tryon calls the 'domain validity coefficients'—to be determined mathematically.

Further work in support of the Shevky concept has been done by van Arsdol, Camilleri and Schmid.[1] Rather than looking at individual cities, they have attempted to examine a limited model of urban society by studying the combined results of 10 large American cities. Taking census tract measures on Shevky's six variables (in other words excluding the rental variable which was dropped in later monographs), they examine the patterns of intercorrelations by a factor-analysis interpretation. Their data suggest that Shevky's indexes do differentiate at least three dimensions and that these dimensions are related in the manner specified by Shevky. However, while *social rank* and *urbanization* are suggested as being independent, *segregation* is negatively related to *social rank* and positively related to *urbanization*.

When the data for all ten cities are combined, the pattern of intercorrelations of the six variables does follow the hypothesized structure: in other words occupation and education are highly correlated with the single factor *social rank*; fertility, women working and single-family dwellings are correlated with the single factor *urbanization*; and spatially segregated groups form a relatively separate dimension. However, in looking at these patterns for individual cities, anomalies begin to emerge. In two of the ten cities, for example, it was found that fertility was more closely related to *social rank* than it was to *urbanization*. It was necessary to suggest an alternative model, differing from Shevky's, for these two cities and also for a further two cities which showed high associations between fertility and *social rank*. Yet another alternative model had to be suggested for a fifth city in which the proportion of single-family dwellings showed a high negative correlation with *social rank*. In other words, no fewer than half of the cities did not conform to the predicted patterns upon which Shevky's approach depends even though the aggregate data could be seen as a partial validation of Shevky's ideas.

Some doubt is therefore thrown on Shevky's technique and particularly

---

[1] M. D. van Arsdol, Jr., S. F. Camilleri, and C. F. Schmid, 'The generality of urban social area indexes', *Am. Sociol. Rev.* XXIII (1958), 277–84. See also *idem*, 'An application of the Shevky social area indexes to a model of urban society', *Social Forces*, XXXVII (1958), 26–32.

on the rather amorphous measure of *urbanization*. This doubt occurs even when simply considering the interrelations between the variables which Shevky isolates as being most diagnostic. In considering the whole range of possible parameters which might be selected as measures of aspects of urban social structure one might therefore have even graver doubts as to the validity of isolating those few indexes which Shevky has suggested. Such misgivings have been widely expressed by those critics who have taken issue with the proponents of the Shevky analysis since its introduction. Duncan, for example, has suggested that the technique is simply a classificatory concept which is not tied in with any theoretical formulations.[1]

Indeed, it is hard to avoid the conclusion that there is a strong element of *ex post facto* rationalization in the theoretical exposition which Shevky and Bell provide as a justification of the selection of the variables and indexes which they select as diagnostic measures of urban social structure. Their theoretical background draws upon Colin Clark's writings on the concomitants of economic growth and on Wirth's concept of urbanism. In deriving their three basic indexes, Shevky and Bell start by suggesting three postulates which follow from the increasing scale of society. The consequences of these three postulates are then followed through to justify the selection of the three indexes in the following manner.[2] The first postulate related to increasing scale is a 'change in the range and intensity of relations' which leads to a changing distribution of skills and the lessening importance of manual work. Within a given social system, this leads to changes in the arrangement of occupations and so to the first index of *social rank*. The second postulate is seen as the 'differentiation of function' which is seen in the changing structure of productive activity, the growing importance of cities and the lessening importance of the household as an economic unit. Within a given social structure this leads to the movement of women into urban occupations and the spread of alternative family patterns. In turn, this leads to the second index of *urbanization*. The third postulate is the growing 'complexity of organization' which is manifested in the changing composition of population with its greater movement and increasing diversity. Within a given social structure this produces a redistribution of population in space and the isolation and segregation of groups. This produces the third index of *segregation*.

If each of three steps is followed, it can readily be seen that there is no logical procedure whereby the selection of the particular indexes can be justified. At each of the various stages in the argument there would appear to be a variety of possible derivations in place of those selected by Shevky.

[1] O. D. Duncan, review of Shevky and Bell, *Am. J. Sociol.* LXI (1955), 84–5.
[2] Shevky and Bell, *Social area analysis*, pp. 3–5.

Furthermore, this scale transition in the argument, from the scale of a total society to an intra-urban level, would not appear to be the most fruitful basis on which to provide criteria for the selection of variables which are diagnostic in the analysis of urban social structure.

Hawley and Duncan have added a further criticism that there is a confusion within the writings of the Social Area Analysts between social areas and geographical areas.[1] Nowhere is it made clear what is meant by a 'social area' as distinct from a geographical area. In actual practice, the research which has made use of the typology has continued the tradition of the classical ecologists and has clearly had in mind the delimitation and analysis of geographical areas within cities. One recent study, for example, examines four medium-sized American cities and attempts to relate the spatial distribution of the scores of *social rank* and *urbanization* with the urban models of Burgess and Hoyt, coming to the interesting conclusion that whereas *social rank* is a sectoral feature *urbanization* is a concentric characteristic with high scores in the inner areas and lower scores in peripheral areas.[2] In the initial formulations of Shevky and his associates, however, the concept of the 'social area' is ill-defined and confusing.

The great majority of studies which have attempted to test Shevky's suggestions have therefore to be viewed with certain reservations. Most have been conducted by the proponents of the method itself and have been restricted to cities from the West Coast of America from whence the formulations of the technique derived. Studies which have broadened the geographical area from which case-studies have been drawn have included a number of results which have been inconsistent with Shevky's formulations. And such empirical studies have been conducted against a background of theoretical criticism which has drawn attention to the lack of theory-based justification of the method and to the shortcomings of certain of the constructs which have little empirical justification.

The little work which has been conducted outside North America suggests much more damaging criticism of the Shevky technique. Only two such studies appear to have been undertaken and, in spite of the acceptance of the method by the respective writers, a number of anomalies can be demonstrated within their results. It is as well to consider the results of these studies since they illustrate some of the limitations of the typology as applied to non-American data.

In a study of Rome, McElrath examines the interrelationships between *social rank*, as measured by the proportion of non-manual workers and the

[1] A. H. Hawley and O. D. Duncan, 'Social area analysis: a critical appraisal', *Land Economics*, XXXIII (1957), 337–45.

[2] T. R. Anderson and J. A. Egeland, 'Spatial aspects of social area analysis', *Am. Sociol. Rev.* XXVI (1961), 392–8.

proportion of illiterate persons, and *urbanization*, as measured by the fertility ratio and the proportion of women in the labour force.[1] He thus eliminates the measure of single-family dwellings and replaces education by a measure of literacy but, in essence, he reproduces the two basic constructs which Shevky suggests. In considering the validity of the technique he notes that the correlation matrix shows a high degree of intercorrelation between all four variables rather than simply between the variables forming the two supposedly unrelated indexes. The obvious question which this raises is whether the two constructs of *social rank* and *urbanization* are independent of each other or whether they measure aspects of the same fundamental component of social structure. He asks a rather different question: 'Do urbanization variables improve the relationship between social rank variables and the converse?' As he demonstrates, in each case the zero-order correlation coefficients between occupation and literacy on one hand and fertility and women working on the other are significantly lower than the multiple correlations between these two sets of variables and each of the remaining variables. By adding variables, in other words, the levels of 'explanation' have been increased. As he comments, 'This indicates that urbanization and social rank components jointly account for greater variation than either set of components'.[2] However, this does not answer the more fundamental question of whether the two constructs are independent of each other. McElrath slides over this question by calling to his support the results of studies drawn from entirely different areas: 'Therefore, given these significant, albeit small, differences and given the additional fact that this correlation test holds in the ten cities (of America) and in San Francisco *where independence of dimensions was established by separate analyses*, it clearly must be concluded that urbanization and social rank components cannot be substituted one for another without loss in analytic power' (italics added).[3] Findings from American studies can hardly be regarded as justification for ignoring obvious inconsistencies in European material.

In his second set of tests of the validity of the Shevky indexes, McElrath examines the sets of relationships within a matrix which includes, in addition to the four variables which comprise *social rank* and *urbanization*, additional variables which are hypothesized as being related either to *social rank* or *urbanization*. Four variables are considered to be related to *social rank*: crowding (persons per room), and the proportion of persons engaged in primary, secondary and tertiary industries. Three variables are considered to be related to *urbanization*: the dependent population ratio (the population aged under 6 per 1,000 population of 65 and over), the proportion of old

[1] D. C. McElrath, 'The social areas of Rome: a comparative analysis', *Am. Sociol. Rev.* XXVII (1962), 376–91.   [2] *Ibid.* pp. 381, 382.   [3] *Ibid.* p. 382.

persons, and the mean family size. Evidently, if the tenets of the Shevky technique are to hold true and if McElrath's related variables are to assume the relations which he hypothesizes, one would expect the pattern of inter-correlations to isolate two distinguishable groupings: one based on a close association of non-manual and illiterate and showing ties with the four measures of crowding and occupational structure; the other based on fertility and women working and related to the three variables which measure the demographic characteristics of the population. From a rather vague inspection of the size of coefficients, and noting one or two inconsistencies, McElrath

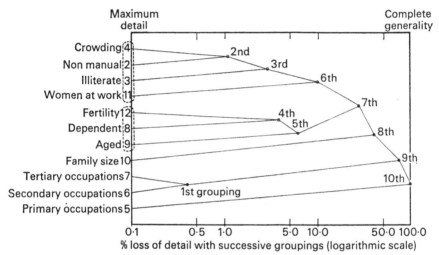

Fig. 2.1. Linkage tree grouping of Rome data

Data taken from D. C. McElrath, *American Sociological Review* (1962). In contrast to McElrath's suggestion that variables 2 to 7 form one group (*social rank*) and variables 8 to 12 form another (*urbanization*), the linkage tree shown here reveals that the most parsimonious grouping is into the groups formed by variables 2, 3, 4 and 11 on one hand and 8, 9 and 12 on the other. Only 10 per cent of the total detail is lost in such a grouping.

concludes that *social rank* and *urbanization* are both necessary measures to describe population characteristics and that they 'effectively describe distinctive population types'.[1] However, if, instead of examining the coefficients with an eye to the theoretical construct of the Shevky technique, an elementary clustering technique is used to make empirical groupings of the most closely related variables in McElrath's set of data, a clearer and very different pattern emerges.

Figure 2.1 shows how, by successively aggregating variables so as to maximize within-group and minimize between-group correlations, a linkage tree can be formed which clusters together those variables which are most

[1] McElrath, *op. cit.* p. 384.

closely related. The objectives and methods of this linkage-tree grouping technique are illustrated by Berry[1] The pattern of the linkages reveals the internal structure of McElrath's matrix. As successive groups are formed the within-group variability increases and the optimal grouping can be determined by the rate of increase of this variability which is shown on one axis of the diagram. After the sixth grouping, with a loss of variability of only 10 per cent, two distinct groups stand out. One is compounded of the following variables: non-manual, crowding, illiterate, and women at work. The other includes three variables measuring the dependent ratio, fertility and old persons. This is a very different pattern from that which would be predicted on the basis of Shevky's technique but the groupings are quite logical. The first group is composed of variables related to socio-economic status, with measures which are familiar parameters in the sociological literature. Most interestingly, it includes women at work and thus suggests that, rather than being an attribute of *urbanization* in Rome, this variable relates to the economic and social position of the household. Women will be found in the labour force not because they are anomic detached units within an urban area, but because their labour is needed to supplement the household income. The second group of variables simply combines a number of elements each of which is related to the age composition of the population. Fertility, like the dependent ratio and the proportion of old persons, is an indirect measure of the age structure of a given population and the grouping of these three variables would appear to be entirely explicable in consequence.

The linkage tree shows that at the seventh combination these two groups are merged, but by doing so the level of internal homogeneity falls rather markedly. Whereas at the sixth combination only 10 per cent of the 'variability' was included in the groups, with the seventh the proportion rises sharply to almost 30 per cent. The most parsimonious grouping would therefore appear to be that outlined above which occurs after the sixth combination. It can also be seen from the linkage tree that the occupational variables are much less relevant within the whole matrix than the remaining variables. The secondary and tertiary occupations are closely related one to the other, but both they and the measure of primary occupations do not combine with the demographic variables until the penultimate and the final combinations respectively. They would therefore not seem to be ideal variables to select as a test of the efficacy of *social rank*.

The utility of the Shevky technique within the context of Rome would therefore appear to be thrown in doubt. A similar conclusion might be made with regard to its use within Britain. Once again, only a single study within

[1] B. J. L. Berry, 'A method of deriving multi-factor uniform regions', *Przegl. Geogr.* XXXII (1961), 263–79.

56

this country has used the technique, and in examining its results our conclusions can only be applied to the town which was used as a case-study but, as will be shown later, there are grounds for suggesting that the same conclusions could be applied elsewhere. Like the Rome study Herbert's study of Newcastle-under-Lyme uses four variables.[1] His matrix of correlation coefficients appears to throw even more serious doubt on the meaning and the validity of the *urbanization* index, since he finds a correlation of only 0·080 between fertility and women in the labour force. Both of these measures are in fact more closely related to the two variables which comprise *social rank* than they are to each other. The meaning of a composite measure which groups two unrelated variables to form a single so-called independent index leaves one rather confused. Indeed, while suggesting that his 'social area map is meaningful and accurately differentiates the urban structure of Newcastle', Herbert is forced to explain the distribution of the index of *urbanization* in terms of a separate consideration of each of its two component variables.[2]

To summarize the conflicting empirical results of these studies, the pattern of their correlation matrices can be compared to the 'ideal' pattern which ought to be found if the Shevky technique did apply to different culture worlds. This can be shown as follows:

(a) IDEAL PATTERN

	Occ.	Ed.	Fert.	W.L.F.
Occ.	.	+	o	o
Ed.	+	.	o	o
Fert.	o	o	.	+
W.L.F.	o	o	+	.

(b) TEN AMERICAN CITIES

	Occ.	Ed.	Fert.	W.L.F.
Occ.	—	0·838	0·546	0·162
Ed.	0·838	—	0·289	0·094
Fert.	0·564	0·289	—	0·636
W.L.F.	0·162	0·940	0·636	—

(c) ROME

	Occ.	Ed.	Fert.	W.L.F.
Occ.	—	0·789	0·676	0·640
Ed.	0·789	—	0·744	0·593
Fert.	0·676	0·744	—	0·685
W.L.F.	0·640	0·593	0·685	—

(d) NEWCASTLE

	Occ.	Ed.	Fert.	W.L.F.
Occ.	—	0·812	0·410	0·271
Ed.	0·812	—	0·316	0·301
Fert.	0·410	0·316	—	0·080
W.L.F.	0·271	0·301	0·080	—

(Signs are ignored and only the four common variables used in the three studies are considered. Occ. = occupation; Ed. = education; Fert. = fertility; W.L.F. = women in labour force.)

It is obvious that only the American results conform to the ideal and, even here, as the authors recognize, the measure of fertility is closely associated with occupation and education, rather than being unrelated as the Shevky theory would suggest. In Rome, fertility is more highly related to education than it is to women in the labour force; women in the labour force is in turn almost as highly related to occupation as it is to fertility and it is also

[1] D. T. Herbert, 'Social area analysis: a British study', *Urban Studies*, IV (1967), 41–60.
[2] *Ibid.* pp. 51–5.

closely related to education. In Newcastle, both fertility and women in the labour force are more closely related to the *social rank* variables than to each other.

These contradictory findings appear to be sufficiently damaging to the concepts of the Shevky typology for its use to be questioned. In the studies which have been considered, the discovery of discrepancies has not prevented the respective authors from continuing with their use of the technique. But what is evidently needed, given the desirability of a technique which can reveal aspects of urban social structure and which delineates urban sub-areas, is a method which is not distorted by the dictates of a possibly suspect theoretical foundation: a method which can empirically select diagnostic variables which can be shown to be orthogonal (that is, unrelated) or whose associations with each other can be measured and taken into account.

MULTIVARIATE ANALYSIS

The Shevky technique might be thought of as a type of multivariate analysis in that it hypothesizes that three factors can be compounded of a number of separate measures, and are necessary to explain the total variation within urban social structure. Most of the criticism which has been levelled at the technique relates to the process by which these factors have been derived. The objective may be sound even though the method has been found suspect. To work towards the same end as the Shevky analysis while avoiding the weakness of selecting variables on the basis of predetermined deductive theory, the objective statistical means of multivariate analysis obviously meet our requirements. Whereas the Shevky technique selects its constructs, and the variables which compose them, on the basis of possibly suspect theory, multivariate analysis selects its discriminating factors solely on the basis of the intercorrelations of the data itself—and a large body of data at that.

Factor analysis and the related technique of component analysis are methods of discovering the structure of a multivariate universe of data by revealing the clusters or bundles of closely related elements contained within a matrix of correlation coefficients. The variation between individuals, which is originally expressed in terms of scores observed on a number of variables, is re-expressed in terms of a number of secondary or derived variables which are compounds of the initial data. These 'components' or 'factors' (when the axes between factors are kept orthogonal) will be uncorrelated one with another. The technique therefore meets our need to isolate those elements within urban social structure which can be thought of as 'fundamental' in that they account for a large fraction of the total variation within the universe of data. The fundamental components, in other words, are comparable to the role which, as was noted above, income played in the three-variable universe of income, rental and the sex ratio.[1]

In component analysis, the covariance or correlation matrix is broken down into a new set of orthogonal components or axes which is equal in number to the original number of variables. The components are weighted sums, or linear combinations, of the original data. The latent roots of this derived matrix of components show the proportion of the total variation which is accounted for by each component. The latent vectors show the loadings of each original variable on the component in question.[2] The loadings thus reveal the type of variation which is contributed by each component. Although p variables will therefore produce p components, the components are usually extracted in descending order of magnitude and, usually, only a relatively small number of components is needed to account for a large proportion of the total variation. For example, Stone has reduced a total universe of seventeen economic variables to three components which account for 97·4 per cent of the total variation.[3] In this way, component analysis aims to reduce a large fraction of the total variation expressed in all the variables by a relatively small number of components which are uncorrelated with each other. The components have high intra-correlation and low intercorrelation and so meet the need for extracting bundles of interrelated elements within the whole complex of urban social structure.

The basic difference between component analysis and factor analysis is that whereas component analysis transforms the original variables into an equal number of uncorrelated components, only a few of which may be needed to summarize the total variation, factor analysis starts from a different premise, namely that, given p variables which form the original observations, the total variation can be expressed in terms of k factors (plus residual error elements) where k is smaller than p. The objective is therefore to express the covariation in as few as possible factors with as small a residual error as possible. To do this implies that the factors which are isolated can be interpreted meaningfully, and this has led to the practice of rotating the factor axes so that variables of a particular type have high loadings on a given factor and zero or near-zero loadings on other factors. Where axes are rotated in this way, the factors are not necessarily kept orthogonal, and this has been one source of criticism of the method since the lack of uniqueness of the axes has been seen as making the technique too subjective. A further difference

---

[1] See above, p. 46.

[2] The terms 'latent root' and 'latent vector' are respectively interchangeable with the terms 'eigenvalue' and 'eigenvector' which are more commonly used by American writers.

[3] J. R. N. Stone, 'The interdependence of blocks of transactions', *J. R. statist. Soc. supplement*, IX (1947), 1–32.

between component analysis and factor analysis is that in factor analysis, since the original variation on the p variables is expressed in terms of k factors (where k is smaller than p), an underlying assumption is that only a part of the unities on the diagonal of the correlation matrix is due to the k factors. Where k is smaller than p the unities are therefore replaced, usually by estimates of the amount of the unities which is accounted for by the k factors. In practice, the unities are often replaced by the highest correlation coefficient of each individual variable.

These differences should not obscure the essential similarity between the two methods. Both are powerful tools in the analysis of a multivariate universe of data. They recognize the fact that in a collection of variables, each individual variable is not of the same importance or weight as a diagnostic measure of the total variation, and that some of the variables overlap to show the same basic patterns of variation. In such a situation, the two statistical techniques suggest which are the diagnostic and which are the redundant variables and isolate the basic patterns that lie within the data. The factors or components are approximations of these basic patterns. The utility of the technique can be seen from its geographical use, for example, by Berry, who has studied economic development on a world scale by examining 43 indexes of development for a total of 95 countries.[1] Factor analysis reduced these indexes to five basic patterns which accounted for 94 per cent of the total sum of squares of the rank-order correlation matrix between the countries. The value of this, as with multivariate analysis in general, is twofold. First, the composition of the factors or components themselves reveals a great deal about the structuring and interplay of the original universe of data. Allowing for the fact that the data which are fed into the process determine, to a very large extent, the factors or components which emerge, the fact that in Berry's study one factor isolated variables which reflected degrees of technological development, whereas another (unrelated) factor isolated variables relating to demographic elements, suggests that these may be unidimensional axes which have to be taken into separate account in considering the complex phenomenon called 'economic development'. Secondly, by providing values or scores on the various axes which emerge, the technique enables one to regionalize or classify the units of analysis in terms of these axes. Berry, for example, shows the relative positions of the selected countries in relation to the first two of his factors, which together account for no less than 88 per cent of the original sum of squares. In relation to this distribution, he is then able to allocate individual countries to groups by using discriminant functions. Multivariate analysis can therefore be used both in the

[1] B. J. L. Berry, 'An inductive approach to the regionalization of economic development', in N. Ginsburg (ed.), *Essays on geography and economic development* (Chicago, 1960), pp. 78–107.

analysis of the structuring of a set of variables and in providing an areal frame-work of units which incorporates a large amount of data and collapses it into more easily mapped multi-component indexes.

Inter-city study

The value of such a technique to the analysis of urban social structure is evident. The surprising thing is that greater use has not been made of it in view of the fact that at least one study demonstrated the geographical application of the technique as much as thirty and more years ago.[1] Multivariate analysis was first developed and applied to practical research in the field of psychology. Its development owes a great deal to Thurstone who used factor analysis to study basic psychological syndromes.[2] The first fundamental statement of component analysis, again in the field of psychology, was by Hotelling.[3] A suggestion that the techniques might be applied to the analysis of urban sub-areas was made as early as 1941 by Hagood who made use of first-factor scores in the delimitation of agricultural regions.[4] Since this suggestion, the number of studies of urban areas using multivariate analysis is only now beginning to grow.

A number of works have used whole cities as their basic areal unit. Price, for example, has examined the covariation of 15 variables for 93 American cities with 1930 populations over 100,000.[5] Using rotated factors, he extracted four factors before the residuals became too small to work with further. Price identified these four factors as being related to the degree of maturity of cities, the extent of service functions within them, the level of living within them and their *per capita* trade volume. This study was regarded by Price simply as an illustrative example of the use of the technique as applied to city data, and he recognized that whether or not the factors which he identified could be considered as fundamental variables could only be seen by subsequent work which tested their invariance.

At least three major surveys of national city characteristics have used multivariate analysis; one is by Moser and Scott who use component analysis in the study of British towns; a second is by Hadden and Borgatta who use factor analysis to study American cities; the third is by Ahmad in a study of Indian cities. Moser and Scott's stimulating work was the first study to appear

[1] M. G. Kendall, 'The geographical distribution of crop productivity in England', *J. R. Statist. Soc.* A, CII (1939), 21–62.
[2] L. L. Thurstone, 'Multiple factor analysis', *Psychol Rev.* XXXVIII (1931), 406–27.
[3] H. Hotelling, 'Analysis of a complex of statistical variables into principal components', *J. Educ. Psychol.* XXIV (1933), 417–44, and 498–520.
[4] M. J. Hagood, D. Danilevsky and C. Blum, 'An examination of the use of factor analysis in the problem of subregional delineation', *Rural Sociology*, VI (1941), 216–33.
[5] Price, *Social Forces*, XX (1942).

in this country which applied component analysis to the study of cities.[1] Starting from the correlation matrix of 57 social and economic variables for the 157 towns of over 50,000 population in England and Wales, they subsumed 69 per cent of the total variance in their first six principal components and, concentrating on the first four components, which together accounted for 60 per cent of the variance, identified these as follows: the first component was associated with social class differentiation; the second was related to population growth; the third was most closely associated with two groups of variables measuring housing and population changes after 1951; the fourth was strongly associated with housing conditions and especially with measures of overcrowding. They note that social class emerges as a discriminating factor in urban development and that the developmental changes grouped in the second and third components are closely associated with such demographic characteristics as age structure and birthrates. Using the scores of towns on the first four components, they are then able to develop a classification of British towns based on between-group and within-group generalized distances in the four independent dimensions determined by the components. Their evaluation of this classification is modest, since they recognize that while it may be of some relevance to many aspects of urban structure it does not reproduce with exactness any single feature of urban differentiation. The multi-dimensional classification 'is meant to be an average classification, and it has the advantages and disadvantages of any average'.[2]

Hadden and Borgatta claim rather more for their study.[3] Their use of factor analysis was extended to cover eight sets of data for different types of city units: towns, and small, intermediate and large cities covered aspects of size; central cities, suburbs and independent cities covered aspects relating to differential relationship to the system of metropolitan dominance; and a final set of data subsumed all the cities together. These eight parallel sets of data therefore enable comparisons between different types of urban groupings. A total of 65 variables are incorporated which cover demographic, housing and economic data and they claim that the parallel studies reveal a great deal of comparability as between each of the eight sets. In all, eight principal factors are named and these are used to isolate diagnostic indexes. In lengthy tables, covering each of the 674 American cities of over 25,000 population, they present city profiles on the basis of 12 of these diagnostic indexes. Having devoted the first part of their work to a bitter and at times repetitive criticism of urban classifications as ends in themselves, this may appear a

[1] C. A. Moser and W. Scott, *British towns : a statistical study of their social and economic differences* (London, 1961).
[2] *Ibid.* p. 93.
[3] J. K. Hadden and E. F. Borgatta, *American cities : their social characteristics* (Chicago, 1965).

somewhat odd end-product, even though they claim stability for the factors which they enumerate.

The third study of national city characteristics is the most geographical in its approach. Ahmad studies 62 variables for 102 Indian cities and reproduces almost 70 per cent of the total variability in terms of nine well-defined principal components.[1] These are shown to have some consistency in that separate analyses, using alternative areal definitions of his urban units and a reduced number of variables, reveal approximately the same composition of components. Using the scores for these ten components, Ahmad then applies grouping techniques which produce classifications of the cities. These groupings show very strong regional elements and consequently are able to form the basis of a regional pattern of systematic spatial variations. The Indian cities are shown to fall into five major groupings which consist of the metropolitan centres, the suburban cities around Calcutta, and three regional groupings of northern, southern and central cities.

Other studies using multivariate analysis have concentrated on various other aspects of city life. Hofstaetter, for example, has followed up Thorndike's measurements of the 'goodness' of a city by studying the factorial structure of variables relating to diseases, public services, literacy and so forth.[2] His list of rotated factors include such fancifully titled constructs as 'enlightened affluence'. This illustrates one of the most important aspects of multivariate analysis: that what goes in determines what comes out. As many of the scholars who have used the technique have pointed out, multivariate analysis is no conjuring trick. The selection of original data must, to a greater or lesser extent, control the 'bundles' which emerge.[3] If social parameters alone are used as the initial observation, one can expect that a social component will emerge as a highly diagnostic factor. Hadden and Borgatta, for example, deliberately duplicated a number of economic variables in their universe so as to force the occurrence of factors which corresponded to some *ad hoc* economic specialization types.[4] While it is interesting to note that, in spite of this, such economic factors were of little overall importance compared to the demographic factors, the fact that these economic factors did emerge illustrates the importance of the input data. The conclusion must be that any attempt to test for the stability of factors or components can only be achieved, either by comparison of a given area at two points in time or in two different regions at one point of time, so long as

[1] Q. Ahmad, *Indian cities: characteristics and correlates*, University of Chicago, Department of Geography Research Paper no. 102 (Chicago, 1965).
[2] P. R. Hofstaetter, '"Your City" Revisited: a factorial study of cultural patterns', *Am. Catholic Sociol. Rev.* XIII (1952), 159–68: E. L. Thorndike, *Your City* (New York, 1939).
[3] A caveat which is strongly emphasized in E. Gittus, 'Statistical methods in regional analysis', University of Strathclyde, *Regional Studies Group Bulletin*, no. 3 (1966), p. 5.
[4] Hadden and Borgatta, *American cities*, pp. 56 ff.

the same set of initial variables is used in the analysis. Unless the technique is used as a classificatory device which has some further end-product in mind, the classification of units can only be regarded as stable on the basis of further testing in different areas or at different time periods. Moser and Scott appear to approach their results in this respect with a good deal more healthy scepticism than do Hadden and Borgatta.

Intra-city study

When we turn from multivariate studies using whole cities as their basic units, to studies which use contiguous urban sub-areas as their basic units, this problem of classification as an end in itself appears less serious. Multivariate techniques are here more often used as techniques of regionalization which, in addition to producing regional groupings, also reveal something of the internal structure of the sets of variables under consideration. Thompson *et al.* have, for example, illustrated the utility of the techniques as a means of studying the economic structuring of New York State, and Carey has examined the structure of Manhattan's demographic and housing patterns using the distribution of six factors which are identified with varying degrees of precision.[1]

Within this country, the release of census enumeration district data has only recently made such studies possible, and from the few which have been completed some interesting parallels emerge which suggest the usefulness of the technique. The study of Sunderland, which is discussed in more detail later, was prompted by the successful application of component analysis by Moser and Scott to the towns of England and Wales. At the same time a number of other studies were being conducted by members of the Inter-University Census Tract Committee.[2] To date, the results of only three completed studies are available and a comparison between two of them furnishes us with an example of the use of component analysis in the study of urban social structure. The major study is Gittus' work on Hampshire and Merseyside, which covers a much greater areal extent than the Sunderland study.[3] Whereas in Sunderland the total number of enumeration districts was 263, Gittus' work included 1,800 in Merseyside and 1,700 in Hampshire. Yet while the scale of the studies is different, there is some similarity between the sets of variables which were included. Done independently, the two

[1] J. H. Thompson, S. C. Sufrin, P. R. Gould and M. A. Buck, 'Toward a geography of economic health', *Ann. Ass. Am. Geogr.* LII (1962), 1–20; G. W. Carey, 'The regional interpretation of Manhattan population and housing patterns through factor analysis', *Geogr. Rev.* LVI (1966), 551–69.   [2] See above, p. 40 n.
[3]. E. Gittus, 'An experiment in the definition of urban sub-areas', *Trans. Bartlett Soc.* University College of London, XI (1964–5), 109–35.

64

studies unfortunately include a different mix of variables, however many of them do overlap. Both sets of variables, 30 in the case of Sunderland and 31 in the case of Merseyside and Hampshire, include a cross-section of demographic and housing parameters, and at least 15 of the selected variables are closely parallel in the two studies, either being identical or virtually identical measures of particular characteristics. The main areas of difference are, first, that the Gittus study included information on place of birth and included more detailed information on types of buildings and secondly, that the Sunderland study included data drawn from the 1961 census 10 per cent sample (relating primarily to social class and education) and also included data drawn from non-census sources (rateable value records and the register of electors). Discussion of the use of 10 per cent data is deferred until the more detailed examination of the Sunderland results.[1] The main point is that both studies included in their list of variables strong emphasis on census material relating to household composition, tenure and density and to housing characteristics.

In view of this similarity, a comparison of the results of the studies seems in order, but since the data were by no means strictly comparable, only the main points of similarity will be noted. First it is of interest to note the closeness between the general amounts of explanation afforded by the components which were isolated by the studies. Gittus found that her first five principal components accounted for 65 and 69 per cent of the total variability in Hampshire and Merseyside respectively. In Sunderland the comparable figure was 77 per cent. Distributed between the components, the amounts were as follows:

Percentage explanation of total variability

	Components					
	1	2	3	4	5	Total (1–5)
Hampshire	26	15	11	8	6	65
Merseyside	26	22	9	7	5	69
Sunderland	30	29	8	6	4	77

In other words, the first two components account for a relatively high degree of the total variability with a marked fall to the third and subsequent components. In all three cases, two-thirds or more of the total variability in the original data is subsumed under five composite clusters.

While a detailed discussion of the content of the Sunderland components is deferred until a later chapter, some indication of the approximate similarity between the highly discriminating variables in both the Sunderland and the

[1] See below, pp. 158–9.

Gittus studies can be noted here since the points of similarity make an encouraging finding which suggests the usefulness of the technique. While it is not profitable to make any direct comparisons of the content of the individual components because of the use of different sets of variables, and especially the absence of social-class variables in the Gittus study, it is worthy of note that in both studies the variable persons per room was individually the most diagnostic. In Merseyside, it alone accounted for 94 per cent of the first component: in Sunderland it accounted for 85 per cent of the first component. By any reckoning, this would appear to be one of the most significant variables which could be taken into account in considering the structuring of towns such as these. Other discriminating variables might be noted. Such variables should be selected in terms of their closeness with one of the principal components and in terms of the amount of individual variance which is absorbed by the analysis. In the following table are shown those variables which Gittus isolates as diagnostic within her study, and the Sunderland data are shown for comparative purposes:

Percentage of total variability accounted for in five components

	Persons per room	Shared households	Type III[1] buildings	Exclusive use of four amenities
Hampshire	22	13	8	10
Merseyside	25	19	14	11
Sunderland	26	17	—	25

In Gittus' analysis, these four variables were closely associated with one or another of the first four components: in Merseyside, for example, persons per room was the variable most closely associated with component 1; shared buildings was that most closely associated with component 2; type III buildings was that most closely associated with component 3; households with all four amenities was the second closest variable in component 4. Likewise in Sunderland, while type III buildings was not included as a variable, the remaining three were each closely associated with different components: persons per room was closest to component 1; shared households was second highest in component 3; use of four amenities was highest in component 2. Some of these individual variables are moderately highly interrelated, but the fact that certain recurring variables stand out as highly diagnostic raises the question of why they should be of such importance and also of whether, in view of the complexity of the calculations involved in multivariate analysis, it might ultimately be possible to arrive at some set of discriminating variables which could profitably be used as single indexes for studies where time

[1] 'Type III buildings' are those which are only partly residential. See *Census 1961*, General Explanatory Notes, p. x.

and resources forbid a full component analysis. Before such a question could be answered, comparative studies are needed which include a common list of variables. The results of the completed studies, however, do suggest grounds for hope in this direction and also provide the material on which the selection of promising sets of variables could be made. The work of members of the Inter-University Census Tract Committee has attempted to move in the direction of compiling suggested variables which could be used in such comparative studies.

The third study which has been completed to date is that of the London Administrative County by scholars at the Centre for Urban Studies.[1] The components which were extracted in this case differed somewhat from those of the Gittus and Sunderland studies. The first component, for example, isolates high proportions of single persons, many being foreign-born, in small households in furnished rented accommodation and with high occupational and educational qualifications. The second component isolates housing conditions. The third measures high proportions of owner-occupiers living at low densities. The fourth groups large proportions of old people living in small households. The importance of the first component, which might be regarded as a high-status rooming-house syndrome, can be seen as a consequence of the particular concentration of such types of area within London and, indeed, it might be expected that the results of an analysis of this area might differ from those previously noted. The area chosen incorporates only the inner core of the much larger functional area of London. To have achieved stricter comparability, a much greater area would have had to be included. Indeed, to achieve comparative studies not only would identical variables have to be used in subsequent studies, but also the boundaries of the areas selected would have to aim at some type of functional comparability so as to include the whole or comparable parts of a total functional 'city-region'.

THE FORMATION OF SUB-AREAS

So far, only the first contribution of component analysis has been discussed. This is its ability to reveal the underlying patterns behind the complex covariation of a large universe of data and the isolation of variables which are most diagnostic in 'explaining' the total variation. In a purely empirical fashion, it is the aim of such study to lay the groundwork upon which a more powerful theory might be developed. The second contribution of the technique is its ability to provide a grouping of sub-areas which can serve

[1] Centre for Urban Studies, University College, London, 'A note on the principal component analysis of 1961 e.d. data for London Administrative County', mimeographed report, no date.

as an areal framework within which more detailed research can be conducted. Having calculated the eigenvectors appropriate to each variable for each of the principal components, these can then be used as weights which are applied to the original data in standardized form. This provides, for each of the areal units, a score appropriate to each of the components. Thus, in the case of the thirty variables used in the Sunderland study, the component analysis collapses the thirty individual scores for each enumeration district into a mere four composite scores, one for each of the four principal components. Thus, thirty parameters are reduced to four, which summarize 63 per cent of the original detail. In other words, thirty maps can be replaced by four maps in a description of the spatial distribution of the thirty variables. This represents a powerful partial solution to the problem of multi-component mapping and the production of homogeneous areal groupings on the basis of a large universe of information.

The large literature on multi-factor mapping suggests both the extent of this problem and the variety of types of solutions which have been suggested. Such concern is not surprising in view of the fact that areal analysis is at the core of much geographical concern. There are two separate yet related aspects of the grouping of sub-units: one being classification and the other the grouping of contiguous areas.

Even when the process of collapsing a universe of data into a smaller number of factors or components has been achieved, in trying to use these scores in the production of classes or types there still remains the problem of combining units in terms of more than one dimension. In the Sunderland case, for example, there are four principal components which have to be taken into account. How can these four sets of scores for each enumeration district be combined to produce homogeneous regions within the town?

The classification methods attempt to group like areas so as to maximize within-group homogeneity and between-group heterogeneity. Given the type of data which is produced by component analysis the first step might be to produce generalized distances on the four dimensions specified by the four component scores. The Euclidian distance is represented by the D^2 statistic which is a measure of generalized distance. Given n dimensions, the generalized distance is $\sum_{i=1}^{N}(x_i - y_i)^2$, where x and y are areal units and i (1, 2, ... n) is an observation for that unit.[1] In the present case, we would therefore have a generalized measure on four sets of scores for each area, thus reducing a four dimensional space to a single dimension—the D^2 score. With a matrix of such scores it would then be possible to use one of a variety of types of classification procedures, such as the B-coefficient, elementary

[1] See, for example, M. Jammer, *Concepts of Space* (Cambridge, Mass., 1954).

68

linkage analysis, or a linkage-tree grouping procedure such as that used above in studying McElrath's data.[1]

All these techniques are essentially classification techniques which pay no regard to the locational characteristics of the areas concerned. They therefore produce type groups analogous to the species of fauna or flora. The results of the Sunderland study were in fact converted to a D^2 matrix and such classificatory tests were applied to it to produce classes of sub-areas. But, in grouping sub-areas into larger aggregates, there are two distinct objectives which differ one from another and which in consequence require different approaches. Given a set of areal sub-units such as in the diagram below, one may need to produce a taxonomy of like sub-areas in which contiguity is not essential to the grouping procedure. If, for example, the sub-areas are various types of shopping centre, primary interest may centre on the fact that area a may be most similar to h, or that b may be most similar to j. However, if the objective is to build up areas which can be considered as formal or 'homogeneous' regions, contiguity is essential to the grouping procedure and the locational characteristics of each sub-unit are important. The important question would therefore be, for example, which of its adjacent sub-units is area a most like. Only in such a way could homogeneous regions be built up.

a	b	c	d
e	f	g	h
i	j	k	l

It is this difference between classification and regionalization which makes the formation of regional groupings peculiarly difficult. There appear to be few objective tests available which hold out much promise of solving the problems of grouping areal sub-units into contiguous wholes. Some interesting attempts have been made, but each has involved numerous comparisons

[1] For the *B*-coefficient see K. J. Holzinger and H. H. Harman, *Factor Analysis* (Chicago, 1941), pp. 23–34. For its use in geographical work see R. C. Mayfield, 'Conformations of service and retail activities', in K. Norborg (ed.), *Proceedings of the I.G.U. Symposium in Urban Geography, Lund 1960* (Lund, 1962), pp. 77–89. For linkage analysis see L. L. McQuitty, 'Elementary linkage analysis for isolating orthogonal and oblique types and typal relevancies', *Q. J. Educ. Psychol. Measmt*, XVII (1957), 207–29. For the linkage tree procedure see Berry, 'A method of deriving multi-factor uniform regions' *Przegl. Geogr.* (1961); for applications see Q. Ahmad, *Indian Cities*, ch. 3, and R. J. Johnston, 'Multi-variate Regions: a further approach', *Professional Geographer*, XVII (1965), 9–12.

The application and geographic relevance of certain of these techniques, and of classification methods generally, are well discussed in P. Haggett, *Location analysis in human geography* (London, 1965), pp. 254–62, and in B. J. L. Berry, 'A note concerning methods of classification', *Ann. Ass. Am. Geogr.* XLVIII (1958), 300–3.

of adjacent areas. Two main types of statistical techniques have been used. One of these is variance analysis. Since variance measures the total sums of squares of deviations from the mean, it provides a measure of homogeneity for grouped units. By examining the change in the within-group variance which results from adding new units to the group, decisions as to which group fresh units should be added to can be facilitated. Bogue, for example, used a rough version of such a method in forming his state economic areas.[1] Zobler has demonstrated its use more stringently by comparing the ratio of within-group variance to between-group variance in an example in which a decision is made as to whether to group the state of West Virginia with one of three alternative adjacent regions.[2]

A second technique has been the use of correlation analysis, which has best been illustrated in the work of Hagood, a writer who has contributed a great deal to the methodology of regional analysis even though little notice appears to have been paid to her ideas in the geographical literature. In an attempt to form regional groupings of American states based on a total of 102 agricultural and population variables, she collapsed the data into a composite agricultural-population index with weights determined by a principal-components solution.[3] In addition, she compiled correlation coefficients between all pairs of adjacent states with respect to their component scores on the composite index. Thus she produced a state map which included not only each state's composite index score, but also the correlation coefficients between pairs of adjacent, or closely located, states. On the basis of these two parameters she was then able to aggregate states into regions by selecting nodes of closely similar states and adding adjacent sets of states which had both similar composite scores and as high intercorrelations as possible. Evidently, as Hagood herself points out, such a method does not produce a uniquely determined delineation of regions, and a number of cases call for personal judgement as to the optimum solution, but the use of the coefficients does provide a yardstick which helps to formulate decisions with greater precision.

An even more interesting regionalization procedure is introduced elsewhere by Hagood, in which she actually incorporates a measure of location into a factor analysis of data.[4] Taking three variables for the 88 counties of Ohio, she added two locational coordinate variables: namely a horizontal and

[1] D. J. Bogue, *State economic areas* (Bureau of the Census, Washington, D.C., 1951).
[2] L. Zobler, 'Decision making in regional construction', *Ann. Ass. Am. Geogr.* XLVIII (1958), 140-8.
[3] M. J. Hagood, 'Statistical methods for delineation of regions applied to data on agriculture and population', *Social Forces*, XXI (1943), 288-97.
[4] Hagood, Danilevsky and Blum, 'An examination of the use of factor analysis in the problem of subregional delineation', *Rural Sociology*, VI (1941), 216-33.

a vertical measure from a given point of origin. These five variables were then used to produce a composite area index on the first-factor loading of the covariance matrix. The range of values was then split into four groups which, with the exception of seven of the counties, contained contiguous counties. The function of the coordinate measure was therefore to impose contiguity on the grouping procedure. The part of judgement as differentiated from objective methods in achieving homogeneous regions was therefore reduced 'to decisions as to the number of subregions desired and the place where the breaks are to be made in the series of index values in the process of group-ing'.[1] The obvious limitation of such a method, however, is that the greater the number of non-locational characteristics which is included, the weaker will be the cohesive function of the spatial measures and the less the likeli-hood of achieving contiguity. Where distinctly homogeneous regions do not exist, contiguity can only be achieved by virtually ignoring the material which is ostensibly used to produce the regions. Hagood, recognizing the problem, illustrates this by performing a further factor analysis on a matrix which includes the two locational measures and only one other variable. This produces regions which, without exception, are contiguous. The choice is therefore between, on the one hand, including a large amount of detail, through the use of a number of non-spatial variables, with less likelihood of achieving contiguity through the use of locational data and, on the other, of ensuring contiguity, but only at the cost of excluding detail. Nevertheless, the inclusion of certain aspects of the locational characteristics of areal sub-units is a most interesting possibility for regionalizing techniques.

A recent monograph by Berry demonstrates this use of spatial variables as an aid to the formation of regions.[2] In examining the spatial structure of the Indian economy, he performs four factor analyses on a wide variety of material. Each analysis incorporates spatial variables which include measures of latitude and longitude, as well as area, population, and population potential. The use of such variables doubtless has the effect of increasing the likelihood of contiguous areas having similar scores, while the number of non-spatial variables prevents the formation of spurious homogeneity where none exists.

None of these methods would appear to produce unambiguous regions. Rather, they provide evidence which can help in the decision as to which units to group into larger aggregates. Where one is dealing with a very large number of areal units, in which the number of possible combinations is very great, it would appear that the simpler techniques of cross–classification of

[1] *Ibid.* p. 227.
[2] B. J. L. Berry, *Essays on commodity flows and the spatial structure of the Indian economy*, University of Chicago, Department of Geography Research Paper no. 111 (Chicago, 1966).

71

sets of data might be just as effective a means of recognizing areas of relative homogeneity. In the Sunderland case, with 263 areal units and four measures for each area, the most satisfactory method appeared to be just such a simple cross-classification of the four sets of data so that the decision became one of deciding where to make the breaks in the series of data.

The important point is that the component analysis scores provide objective parameters with which sub-regions can be built up and which include the evidence of a large number of variables in collapsed form. Such sub-regions provide a framework within which further research can be conducted. In the present case, the analysis and the regional patterning which emerged from the component analysis were used to select areas within which to examine the development of attitudes to education. In thus providing this essential preliminary step the technique more than justifies itself. It must be stressed, however, that the results of component analysis do not substitute for theory even though they might lead on to the elaboration of theory. Given the complexity of the technique, it is essential that the method does not become an end in itself, but leads on to further work. As Stewart and Warntz have said, 'Mathematical statistics can be a good servant; but only a weak social science, intent on becoming weaker, will welcome it as a master'.[1] To see the value of component analysis in true perspective, its two aspects have to be viewed as steps to the development of a better understanding of social structure. First, the ability of the technique to isolate fundamental patterns and to reveal the most important diagnostic variables provides empirical evidence out of which a more holistic theory of urban social structure might develop. Second, the areal framework which can be built up on the evidence of the spatial pattern of component scores can provide a sampling framework for the selection of meaningful social areas within which hypotheses of structural relations might be tested and theories might evolve.

[1] J. Q. Stewart and W. Warntz, 'Physics of population distribution', *J. Reg. Sci.* 1 (1958), 119.

PART TWO

ANALYSIS

CHAPTER 3

SIMPLE ECOLOGICAL STRUCTURE

In this chapter some of the ecological concepts which have been discussed in the previous section, will be applied to a body of empirical data. The town which is selected as a case study is Sunderland, an industrial town in the North-East of England. Sunderland displays many traits characteristic of the American towns which were the seedbeds of ecological theory, since it is essentially Victorian and the physical outlines of its internal composition are largely a product of nineteenth-century population expansion. In an attempt to delineate the structure of the town, the models of Burgess and Hoyt and the ideas of other ecologists such as Davie will be used to test their validity when applied to an English town. It is suggested that these essentially simplistic models have relevance in the context of nineteenth-century developments, but that they have grave limitations in the analysis of modern developments, and in the following chapter a more powerful type of analysis is developed which goes some way to correcting for these limitations.

CHARACTERISTICS OF SUNDERLAND

Sunderland is a town which is living on the dwindling fat of its Victorian expansion. The legacy of the Industrial Revolution is apparent in its appearance, its industrial structure, its population growth and in a host of social and economic characteristics. Even attitudes are coloured by its past heritage. The Depression years, the final death spasm of the nineteenth-century in a pre-Keynesian era, are still a real memory amongst much of the town's population and impinge upon the attitudes of the working population. This imprint of the past, rooted in a continuing dependence on heavy industry, is found to a much greater degree than in the towns of the Midlands or even Lancashire, since the spread of light manufacturing has had only marginal effects in the North-East.

The growth of the town's population illustrates its nineteenth-century importance. At the 1961 census, Sunderland had a population of 189,686, which represented an increase of only just over 8,000 since 1951, or an average annual increase of only 0·45 per cent as against a national rate of 0·53 per

Urban analysis

cent.[1] Such expansion at rates below the national average is a phenomenon largely of this century since in earlier periods the rates of expansion were often substantially greater than the national averages. Accepting the pre-census population figures, the eighteenth-century growth was considerable: from a figure of 4,000 in 1700, the population rose to just under 25,000 by 1801 (Table 3.1). The nineteenth century continued this expansive trend. From 1801 to 1901 the town showed an almost fivefold increase up to a

TABLE 3.1. *Population growth of Sunderland*

Date	Population	Area	Boundary changes
1481	1,300 (est.)	—	—
1681	3,090 (est.)	—	—
1781	20,940 (est.)	—	—
1801	24,444	Townships of	—
1811	25,180	Bishopwearmouth,	—
1821	30,887	Bishopwearmouth Panns,	—
1831	39,434	Monkwearmouth,	—
1841	51,423	Monkwearmouth Shore and Sunderland Parish	—
1851	63,897	Sunderland M.B.	—
1861	78,211	Sunderland M.B.	—
1871	98,242	Sunderland M.B.	—
1881	116,542	Sunderland M.B.	—
1891	131,015	Sunderland C.B.	1895 (from 3,306 acres to 3,357)
1901	146,077	Sunderland C.B.	—
1911	151,159	Sunderland C.B.	—
1921	159,055	Sunderland C.B.	1928 (from 3,357 acres to 6,305)
1931	185,824	Sunderland C.B.	1935 (from 6,305 acres to 6,938)
1951	181,524	Sunderland C.B.	1951 (from 6,938 acres to 8,575)
1961	189,686	Sunderland C.B.	—

SOURCES: pre-1801: W. Hutchinson, *The History and antiquities of the County Palatine of Durham* (Newcastle, 3 vols. 1785–94). 1801 and subsequent: *Census*.

figure of 146,000. By contrast, the twentieth-century expansion has been very slight, with the population rising by less than one-third between 1901 and 1961. These absolute figures are somewhat complicated by changes in the areal extent of the town and the varying rates of change can be shown more clearly by comparing inter-censal rates for comparable areas wherever this is possible (Table 3.2). These rates show that, in the nineteenth century, there was an irregular fall from the highest rate in the decade 1831–41 to the lowest rate in the decade 1891–1901. The twentieth-century pattern is very different, with markedly lower overall rates of expansion and a tendency for the rates

[1] *Census, 1961*, England and Wales. *County Report : Durham* (H.M.S.O. London, 1963). Hereafter no reference will be given to specific census sources. It may be assumed that census data noted in the text have been taken from one of the census publications for the year in question.

either to remain roughly constant or else to fall. Contraction was most marked in the period 1931–51 when the population showed an absolute decline.

Set within a national context, this pattern is more strongly confirmed (Fig. 3.1). The national figures, compounded of the booms and declines of a large number of towns, show a relatively constant growth rate throughout the nineteenth century and then they fall in the twentieth century to achieve

TABLE 3.2. *Rates of population growth in inter-censal periods*

| Period | Annual average percentage rate of growth | |
	Sunderland	England and Wales
1801–11	0·30	1·40
1811–21	2·27	1·81
1821–31	2·77	1·58
1831–41	3·04	1·43
1841–51	2·43	1·27
1851–61	2·24	1·19
1861–71	2·56	1·32
1871–81	1·86	1·44
1881–91	1·24	1·17
1891–1901	1·09 *i*	1·22
1901–11	0·35	1·09
1911–21	0·52	0·49
1921–31	0·20 *ii*	0·55
1931–51	− 0·27 *iii*	0·47
1951–61	0·45	0·53

SOURCE: *Census.*

NOTES *i* Allowance has been made for 1895 boundary extension by using the 1901 area to calculate the 1891 population.

ii Allowance has been made for 1928 extension by using the 1931 area to calculate the 1921 population.

iii Allowance has been made for 1935 extension by using the 1937 area to calculate the 1931 population. Lack of records (through war damage) does not permit allowance to be made for the 1951 extension, and the 1931–51 fall is therefore almost certainly an understatement of the real decrease.

a second, lower plateau level. The Sunderland rates are well above the national average from 1811 to 1881, but from 1891 to the latest census they are at or well below the national rates. To this extent the end of the century is a turning point in the demographic history of the town, marking the divide between the boom conditions of the middle nineteenth and the stagnation of the twentieth centuries. To a large extent, this difference is the result of emigration rather than a drastic change in the natural population increase of the town, since births have consistently exceeded deaths even in the period 1931–51 and the town has maintained rates of natural increase above the national averages. The heavy loss through outward migration in the twentieth century was at its highest in the war years and in the depression years of 1926–7 and 1933–8. In consequence, until the recent population expansion which produced the 1961 population, the town's highest population has been estimated to have been in 1932. The balance of outward migration in the years between 1895 and 1939 stood at an average annual figure of 1,250 persons.[1]

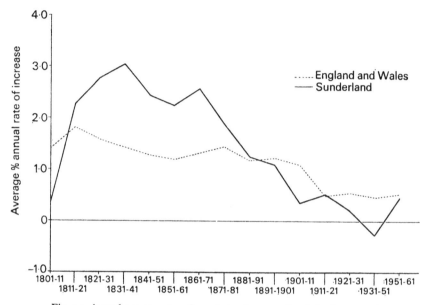

Fig. 3.1. Annual average rates of population increase: Sunderland compared to England and Wales

Source : Census

This pattern of population growth, fluctuation and decline paints a picture of the changing economic fortune of the town. Mitchell, writing in 1919, and looking at the population expansion of the late eighteenth and the nineteenth centuries, lists the factors which he holds responsible for 'this rapid increase in the population and wonderful growth of the town' as:[2]

(i) the appointment of the River Wear Commissioners in 1717,
(ii) the opening of the Wearmouth Bridge which was built in 1796,
(iii) the opening of new collieries in the area,
(iv) the introduction of steam power and railways,

[1] Migration statistics are estimates given in *Written analysis on the survey of the county borough of Sunderland* (Borough Engineer and Surveyor, Sunderland, 1951).
[2] W. C. Mitchell, *A history of Sunderland* (Sunderland, 1919). The sources on which the following nineteenth-century description is based can be found in the bibliography.

(v) the facilities offered for shipping and

(vi) the increased importance of the town after becoming a Parliamentary Borough in 1832.

And it is indeed to the consolidation of the industrial base of the town that one has to look to add substance to the bald figures of population change. It is significant that it was during the inter-censal period of greatest population growth—from 1831 to 1841—that the embryo of the railway pattern of the North-East was being hammered out. As a contemporary of the period commented, 'Yes, truly, nothing, next to religion, is of so much importance as a ready communication'.[1] Throughout the whole area the period was one of fevered economic activity stimulated by the growth and multiplication of railway lines and based largely on the coal-exporting trade and the iron and steel industry which, in the case of Sunderland, principally served the shipbuilding industry. By 1834 as much as 302 miles of track had been commissioned in the North-East, a mileage almost as great as the total laid down by the London–Birmingham, London–Southampton, Grand Junction and Liverpool–Manchester railways which were the principal railways sanctioned up to that time.[2] It was at this time that the competition between the rival ports of Tyneside, Tees-side and Wearside expressed itself in frantic bids to build rival railways to tap the output of the coal mines of the inland areas of Durham. The three towns of Newcastle, Hartlepool and Sunderland in particular were locked in competitive struggle. The opening of the Stanhope and Tyne Railway in 1834 gave the Tyne undisputed pre-eminence in the field of coal exporting and while this line was being built both Newcastle and Hartlepool gave notice of their intention to build lines to tap the rich coal reserves of the mid-Durham area, projects which would have excluded Sunderland from effective competition had it not met strategy with strategy by projecting a rival line from mid-Durham to shipping places on the Wear. Despite the dramatic clash of opposing interests, all three bills were passed by Parliament in 1834 and the Sunderland and Durham Railway began operating in 1837.[3] Such railways consolidated the industrial boom of the whole area and Sunderland ensured that it shared in the expansion. Its population growth at this time reflects the town's economic success and the acumen of its citizens and of entrepreneurs such as George Hudson, the 'Railway King', who laid his Midas touch upon the town.

The connection between population growth and economic vitality is axiomatic. If the multiplication of railways, which began during the decade

---

[1] From a speech in 1841 at the opening of a new railway line from York to Darlington. Quoted in R. S. Lambert, *The railway king, 1800–1871 : a study of George Hudson and the business morals of his time* (London, 1934), pp. 62.

[2] W. W. Tomlinson, *The North-Eastern Railway* (Newcastle, 1914), p. 234.

[3] *Ibid.* p. 233.

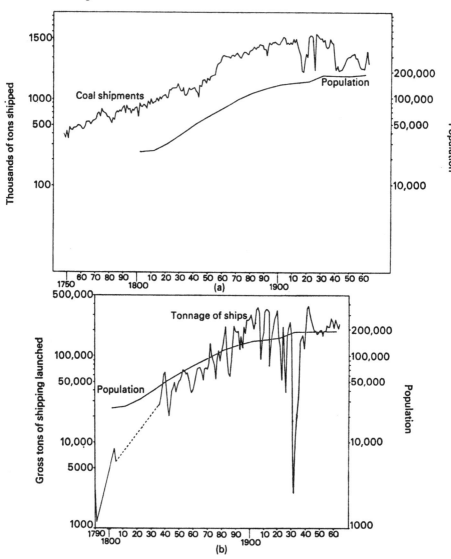

Fig. 3.2. Rates of population growth compared to (*a*) the shipment of coal and coke from Sunderland, and (*b*) the tonnage of shipping launched from the town

Sources: (*a*) J. U. Nef, *The rise of the British coal industry* (London, 1932); G. Garbutt, *A historical and descriptive view of the parishes of Monkwearmouth and Bishopwearmouth and the port and borough of Sunderland* (Sunderland, 1819); River Wear Commissioners, personal communication.

(*b*) J. W. Smith and T. S. Holden, *Where ships are born* (Sunderland, 1953); Wear Shipbuilders' Association, *Annual Reports*

1830–40, laid the basis for the industrial development of the town, the fortunes of the two staple industries of the town fill in the detail of this expansion and provide a rationale for the picture of a booming nineteenth-century provincial town. The population growth which has been examined can be explained by reference to the development of these two activities.

Fig. 3.2 shows that both coal exporting and shipbuilding expanded almost continuously throughout the nineteenth century. As might be expected, shipping tonnages show much greater fluctuation since the individual units of production are relatively large and the launching of a ship at the end of one year rather than at the beginning of another can cause large fluctuations in annual output. To this extent, the shipping returns are a less sensitive indicator of economic conditions than are the coal returns. Nevertheless the rates of expansion are remarkably similar for both. The most interesting feature, however, is the marked parallelism between these rates of growth and the rate of population expansion, during the nineteenth century.

The coal returns consistently expanded throughout the nineteenth century but never at so fast a rate as between 1840 and 1860. This was a period of high population increases which followed immediately on the establishment of an adequate railway system linking the town not only with mid-Durham, but also with areas to the south via the Londonderry line of 1854 and to the north and north-east via the Brandling Junction line of 1839. The output of ships shows only a slightly different pattern of growth with expansion occurring at a higher, although more fluctuating, rate right up to the turn of the century. Nineteenth-century histories of the town give some idea of the industrial unrest and disruption which this rapid expansion caused and at the same time are filled with Victorian wonder at the increase of the size of ships launched and the multiplication of yards, especially in the earlier part of the century before a certain degree of rationalization was produced by the rigours of the market forces.

The main point to bear in mind, however, is the close association between the expansion of all three factors during the nineteenth century namely, population, shipbuilding and coal exporting. In the twentieth century, this close association no longer holds good to the same extent. While, like the growth of population, both shipbuilding and coal exporting reached peaks around the turn of the century, the effects of economic depression, industrial unrest and world wars understandably had a much more direct impact on their subsequent expansion than on the population of the town. The present position mirrors the industrial experience of the country as a whole with declining coal exports and fluctuations in shipbuilding which owe more to the state of world markets than to industrial conditions within the country itself.

Urban analysis

The town's dependence on these two staple industries is highlighted by the general industrial history of Sunderland. Other industries have been of importance but only for relatively brief periods of time. Pottery and glass-making, for example, were of importance during the eighteenth century, but as the nineteenth century progressed, they contracted and finally disappeared so that today there is little trace of the potteries which had lined the river banks, save for the legacy of street names or the examples of their products preserved in the local museum. Similarly, the glass and bottle works are preserved only in the presence today of the single specialist glass works of 'Pyrex'. Other industries recorded in the early histories of the town include salt-works, lime-kilns, copperas works, and roperies. With the exception of the roperies, which form an ancillary industry to shipbuilding, the early importance of such industries has now disappeared.

The continuing dependence upon this limited industrial range is shown by various census returns. In 1861, for example, 42 per cent of the occupied males aged 20 and over were directly engaged in occupations related to ship-building and shipping. Of the shipping employment, an overwhelming pro-portion was concerned with the export of coal either for export abroad or in the coastal traffic to London and the East Coast ports. If to this figure is added that for persons engaged directly in coal mining, one finds that, as far as can be calculated, nearly half of the employed male population was dependent on the two basic industries. Many others were also employed in industries ancillary to shipbuilding. By the end of the century, this depend-ence was still in evidence. While census categories are in no way comparable between one period and another, it can be shown that in 1911, for example, of the occupied males of 10 years and over more than 20 per cent were employed in the 'conveyance of goods, men and messages, on sea, rivers and canals' and in the 'construction of ships and boats'. At the present time, this dependence on a limited range of industries continues as a persistent feature of the occupational structure. Of the total occupied male population aged 15 and over in 1951, 20·6 per cent were engaged in shipbuilding and repairing (Standard Industrial Classification, VI, 50), 5·6 per cent were in marine engineering (S.I.C. VI, 51), 4·1 per cent were in water transport and harbour services (S.I.C. XIX, 224.6), and a further 4·2 per cent were in coal mining (S.I.C. 11.10). Even excluding ancillary industries, over a quarter of the total population was directly involved in the two staple industries.

It is this dependence on a few heavy industries which gives the town much of its present character and gave it character throughout the nineteenth century. The top-heavy industrial structure has had repercussions in a wide range of social and economic fields. Not least of course have been the direct effects on unemployment. Judged by this barometer of economic vitality,

Sunderland is a town of precarious economic health. The White Paper on the North-East has shown the differential between the unemployment figures of the North-East and northern region as against the national figures for the period since 1923.[1] Throughout this period, at no time was the northern figure lower than the national figure. Nor is the pattern of unemployment evenly spread throughout the northern region. Among the larger towns of the area, the major service centres such as Newcastle and Darlington show consistently lower rates of unemployment than the surrounding towns. This is a function of their more broadly based occupational structures which arise from the tertiary industries which they attract and which provide an employment stability conspicuously lacking in the neighbouring towns. Sunderland, for example, has experienced unemployment rates well above even the regional levels. Comparing the national, regional and Sunderland figures, while all three show the same periodic movements and a similar secular trend, the absolute figures for Sunderland rise to three and more times the national figures and show much larger upswings as the economy contracts (Fig. 3.3). This differential always remains; in more prosperous times it tends to decrease, while in poorer years it tends to increase. While one might expect a wider range in these figures since the unit of observation is smaller, the consistency of the pattern, which shows Sunderland at so marked a disadvantage, emphasizes that the distinction is a real one. At the height of the slump of the early 1960s, for example, when the recession in shipbuilding led to the closing of yards and was having a multiplier effect throughout the industries of the town, the Industrial Development Officer noted 1962–3 unemployment figures of as high as 9,000 out of a total insured population of less than 88,000. The figure for June 1963 represented 7·4 per cent of the insured population compared to a regional percentage of 4·3 and a national percentage of only 2·5. 'In effect, Sunderland is still faced with the situation that even if all the jobs expected materialize, and if there is the hoped-for recovery in the shipbuilding industry, there will still be a need to attract, over the next five years, another 5,000 new jobs to secure full employment.'[2]

This picture of economic insecurity has persisted despite the efforts of successive governments since the Distressed Areas legislation of the 1930s. The culmination of this concern was reflected in the publication of White Papers on both the North-East and central Scotland, which recognized the peculiar status of these areas in the national economy and the apparently intractable problems which faced them. The cumulative efforts of government assistance have not been without effect in Sunderland. The Trading Estate

[1] *The North-East : a programme for regional development and growth*, Cmnd. 2206 (London, 1963), pp. 8–9.
[2] Sunderland Corporation, *Annual report of the industrial development officer* (Sunderland, 1963), p. 3.

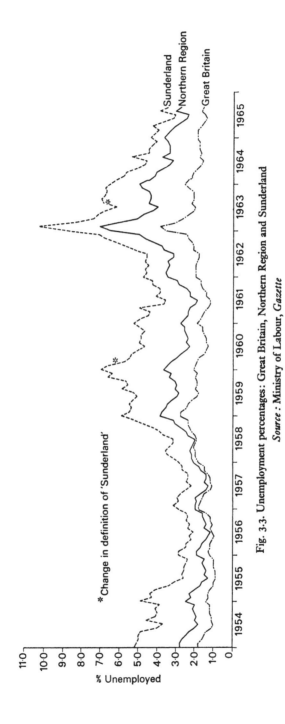

Fig. 3.3. Unemployment percentages: Great Britain, Northern Region and Sunderland

Source: Ministry of Labour, *Gazette*

Movement led to the establishment of the Pallion Trading Estate in 1938 and, in the post-war period, factories both on this estate and elsewhere in the town have helped to do something to diversify the town's industrial structure. Apart from the important historical legacy of the glass works, the light industries which reduce the dependence on shipbuilding and repairing cover a wide range of products. Since 1938 the range of industries has come to include the manufacture of furniture, scientific instruments, cathode ray tubes, aero parts, clothing office equipment, industrial gases, packaging materials and other light manufacturing. The volume of this new light industry has not been negligible but, while it represents an attempt to reduce the over-dependence on a small range of industries, it is questionable whether the basic economic insecurity has been reduced to any considerable degree. As with much industrial decentralization, it has been branch factories rather than parent firms which have moved to the town and it is in such branch factories that contraction tends to be concentrated in times of economic depression. Further, the impact upon the male labour force has been much less than upon the female sector. The semi-skilled work involved in the clothing, television component and the telephone and electrical equipment factories has become overwhelmingly female. The actual extent of light industrial development has also been slight by comparison with that in the metropolitan parts of the country. While the absolute amount of industry has not been negligible, the relative amounts have been much smaller than elsewhere. House, for example, shows that the rate of industrial building in the North-East development area has steadily fallen behind the national rates. The area still has significantly lower proportions of its labour force in the faster growing industries and higher percentages in the faster declining industries by comparison with the national figures.[1]

Set within a national context, the development of light industries in Sunderland can therefore be seen to represent a much less significant development than might at first appear. It is still a town of heavy industry and subject to the fluctuations of economic fortune and the distress and poverty which this implies in the present economic situation. In so far as recent trends have emphasized the emergence of white-collar work and increasing affluence, Sunderland is still something of a Victorian town and much of its ethos, character and structure are a direct product of the forces which moulded it in Victorian times. In their classic comparative work on British towns, Moser and Scott assign Sunderland to a category of industrial towns where it shares a place with Gateshead, South Shields, West Hartlepool, West Ham, Barnsley, West Bromwich, Salford, Warrington, Merthyr

[1] J. W. House, *Recent economic growth in North-East England*, University of Newcastle, Department of Geography Research Series no. 4 (1964), pp. 18–26.

85

Tydfil and Rhondda.[1] The cross-classification of their two principal components places this group of towns in the lower left-hand corner of the two-dimensional space of their graph. One can visualize this two-dimensional space as representing a progression from Victorian to modern with increasing modernity (summarized perhaps as growing white-collar work with all the socio-economic consequences that follow) as one moves higher and to the right of the graph. Sunderland's placing underlines the fact that it is something of a throw-back to the earlier days of the Industrial Revolution. It is therefore intended to conclude this brief summary of the background characteristics of the town by looking at three aspects which, it is argued, illustrate the way in which the nineteenth-century characteristics appear to have persisted through to the present.

First is the relative poverty and low social class of Sunderland. The fact that the town has a generally low social class need not be laboured here except to underline the fact that Sunderland is by no means a one-class town in the sense in which the term has been applied to Dagenham or Bethnal Green. Moser and Scott have shown that Sunderland had 34 per cent of its population in the Registrar General's Social Classes IV and V and that this was the 28th highest proportion out of the towns which were studied. At the same time however, 11·5 per cent of the population was in Social Classes I and II.[2] The town's social hierarchy is therefore bottom-heavy, but not so distorted as to make it unrepresentative of the mass of industrial towns in the country. It is predominantly working class although by no means a one-class town. Understandably, this low social class is reflected in relative poverty. A survey by Comart Research provides some measure of this economic poverty. Of the 201 trading areas into which the United Kingdom is divided by this survey, Sunderland ranks 164th in terms of its index of income per person and 179th in terms of incomes per household.[3] As against an average U.K. personal income of £423, Sunderland's average was £251, and was distributed as follows:

Percentage of Household Income[4]

	Below £500	£500–£1,000	£1,000–£1,500	£1,500 and more
Sunderland	16·0	39·0	27·5	17·5
United Kingdom	13·0	27·7	31·9	27·4

Two other aspects connected with this poverty and low social class are the housing stock of the town and the age structure of its population, which both reflect the perpetuation of nineteenth-century patterns and are worth con-

[1] Moser and Scott, *British Towns*, p. 86.
[2] *Ibid.* pp. 116–17.
[3] Comart Research Ltd, *Survey of incomes and households in the United Kingdom, 1964* (London, 1964).   [4] *Ibid.* pp. 20–1 and 36–7.

sidering in view of the fact that both are used as variables in the multivariate study presented later. The nineteenth-century housing stock was peculiarly poor. Building typically took the form of rows of small cottages similar to the mining villages of interior Durham. At the end of the century, the 1891 census illustrates the high proportion of cramped housing and overcrowding. Over a third of all dwellings were one- or two-roomed as against a national average of 15 per cent. The proportion of persons living at over 2 per room was as high as 33 per cent against a national figure of 11 per cent. While the census notes that the wide regional variation in such figures seems not to have any apparent explanation, it is interesting to note that the three towns with the highest degree of overcrowding were all to be found in the North-East and Sunderland is amongst them. Today, while the absolute figures are much lower, the relative position has remained rather similar. For example, in 1961 Sunderland had the highest average density per room and the fourth highest proportion of persons at over 1½ per room by comparison with all other towns in England and Wales of over 50,000 population:

Average density per room (towns at over 0·80 persons only)		*Percentage persons at over 1½ per room (towns with over 10 per cent)*	
1 Sunderland	0·83	1 Gateshead	16·1
2 Gateshead ⎱ Dagenham ⎰	0·82	2 Willesden	15·2
		3 Newcastle	13·8
3 Huyton with Roby	0·81	4 Sunderland	13·7
4 Willesden	0·80	5 London	11·5
		6 Birmingham	10·7
		7 South Shields	10·1

It is interesting to note how many of the above towns were found in the North-East as a whole.

This is one side of the Victorian legacy. The other side is the energetic response to this legacy of poor housing. Between 1945 and 1964, just under 20,000 houses were built in the town by the local authority. Indeed, the local authority rate of house building per head of population between 1945 and 1958 was the third highest amongst all the 157 towns studied by Moser and Scott. This enormous effort by the local authority is a direct consequence of the poverty of the Victorian housing stock and, while it has made a vast improvement in the living conditions of many of the town's population, the improvement, as has been shown, is absolute but not relative, since rates of overcrowding remain high. Admittedly overcrowding is not necessarily an index of poor physical housing since many of the new council estates are occupied at densities of over 1½ persons per room, yet the high levels of crowding coupled with the remaining high proportions of physically obsolete housing makes it apparent that the town is still much less well provided for in its housing stock than are other British towns.

Urban analysis

The final characteristic to be considered, and again one which reflects the industrial and social composition of the town, is the fertility ratio and the age structure. Differential rates of fertility have long been associated with social class differences even though the pattern appears to be in process of change today.[1] With its high proportion of working-class population, Sunderland, not unexpectedly, has long tended to have high fertility ratios. This has been reflected in the age pyramids of the town, which show throughout the nineteenth and twentieth centuries a significant preponderance of younger persons and a deficit of elderly, compared with Britain as a whole. While the changing pyramids closely follow the national pattern over time, there is a consistent pattern of a broader base and more rapidly declining apex, reflecting the younger age structure of the town. Comparing the percentage excess or deficiency of the Sunderland figures as against the national figure for each age cohort, it is the younger age groups which show an excess in Sunderland throughout the hundred years from 1861 to 1961. The older age groups show a consistent deficit in Sunderland (Fig. 3.4). This pattern is most clearly seen in the female figures where the change, from excess in the younger cohorts to deficiency in the older, occurs within the age ranges from 20 to 34. At no time in the whole of this period did Sunderland have lower proportions of females than England and Wales in any of the age groups below 20. The pattern is slightly less clear in the male figures, but they are similar to the females throughout most of the twentieth century.

THE ASSUMPTIONS OF CLASSICAL ECOLOGY IN RELATION TO SUNDERLAND

The picture therefore emerges, from all this evidence, of a town which is predominantly low class and based to a high degree on a small range of heavy export-orientated industries, notably connected with shipbuilding. This dependence has had diverse effects on the town, but the overriding effect has been to preserve much of what is commonly associated with the Victorian period. Sunderland is indeed something of an historical anachronism not only in the many direct legacies from the nineteenth century, such as the abandoned shipyards along its river banks, but also in the poverty, inadequate buildings, young age structure and bottom-heavy social structure which so characterize the town. To the extent that it was in the nineteenth century that the population was booming and that much of the nineteenth century imprint still remains, Sunderland might be expected to approximate more

---

[1] The relationship of social class to fertility appears to be changing from a negative linear relationship to a U-shaped curve, with large families being found at both ends of the social scale. See, for example, 'Fertility differentials in England and Wales: some facts', *Eugenics Review*, LIX (1967), 70–2.

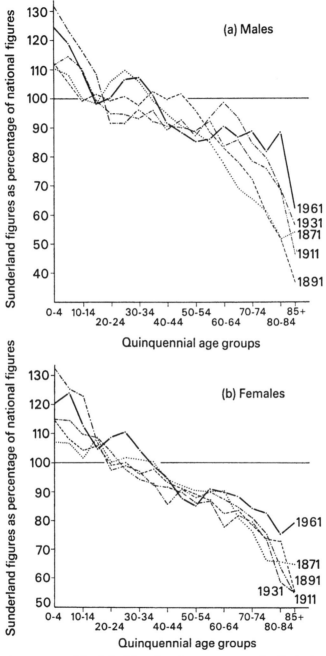

Fig. 3.4. Age structure of Sunderland for selected years as compared to England and Wales

Source : Census

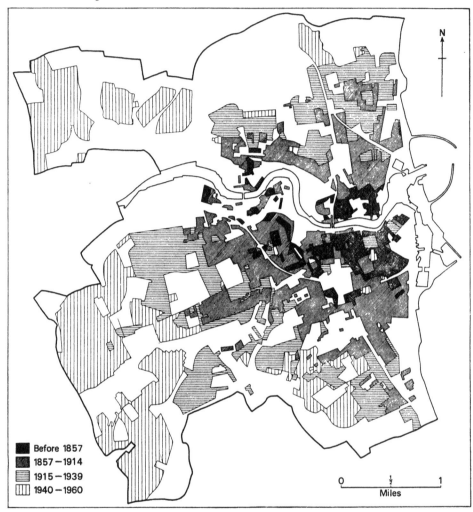

Before 1857
1857–1914
1915–1939
1940–1960

Fig. 3.5. Age of buildings in Sunderland, 1963
Source : Sunderland Planning Office

closely to the assumed pattern of growing towns upon which the early
ecological theories of urban growth were developed. It has been stressed that
one of the assumptions of Burgess and the Chicago ecologists was that the
towns which they studied were undergoing growth. In view of the nature of
Sunderland, it appears to be a peculiarly appropriate town in which to test
some of the ideas of the ecologists, since the tangible remnants of the period
of its greatest population expansion are still the predominant element of much
of its physical structure.

A second assumption underlying the Chicago analyses was that the ideal towns to which their models applied grew outwards in successive rings from a single dominant centre. Before considering the ecological structuring of Sunderland, it would therefore be profitable to see to what extent this second assumption holds true for Sunderland.

In general it appears that, although the present town is formed of the initial fusion of three principal centres, one to the north of the river and two to the south, with the building of the Wearmouth Bridge across the river in 1796, the Central Business District which formed at the southern end of this bridge formed an effective dominant nucleus around which growth has subsequently occurred. The map of the age of buildings shows the manner in which outward growth was orientated about this single core with fresh accretions being added to the periphery of the growing town (Fig. 3.5). The fact that the Central Business District developed to the south of the river and that expansion has tended to be more rapid and extensive in the southern part of the town is doubtless the result of the much more extensive tributary area lying to the south and south-west, since the towns of Tyneside tend to act as service centres for much of the area to the north.[1]

A rather more precise concept of this successive outward growth can be supplied by examination of the population growth within areas of the town. For the nineteenth-century period, the most satisfactory small-area data are provided by the townships, since their boundaries remained fairly constant and where changes did occur adjustments to the figures are possible.[2] Figure 3.6 traces the township changes for a continuous 100-year period and, although some of the townships, particularly Bishopwearmouth, are more extensive than one might wish, the figures provide an interesting picture of the outward expansion of the growing town. The townships fall into three types in terms of varying centrality and varying rates of growth. First are the inner townships of Bishopwearmouth Panns, Sunderland and Monkwearmouth Shore. The first two comprise central areas adjacent to the growing shipyards and industrial areas of the river, and their dramatic decline in population is a consequence of the expansion of this industry and the decay of housing in such inner areas. Both factors led to a centrifugal migration of population as the century wore on. A note in the 1861 census, for example,

---

[1] This is shown by a sphere-of-influence survey in D. A. Burgess, 'Some aspects of the geography of the ports of Sunderland, Seaham and the Hartlepools' (unpublished M.A. thesis, University of Durham, 1961), p. 58. An interesting, but no doubt subsidiary, factor may have been the harsh tenure conditions imposed on building by the landowner who held most of the land lying to the north of the river. See J. Burnett, *History of the town and port of Sunderland and the parishes of Bishopwearmouth and Monkwearmouth* (Sunderland, 1830), pp. 62–3.

[2] The township, or civil parish, was 'an area for which a separate poor rate is or can be made, or for which a separate overseer is or can be appointed' (*Victoria History of the County of Durham*, 3 vols. London, 1905–28, II, 261).

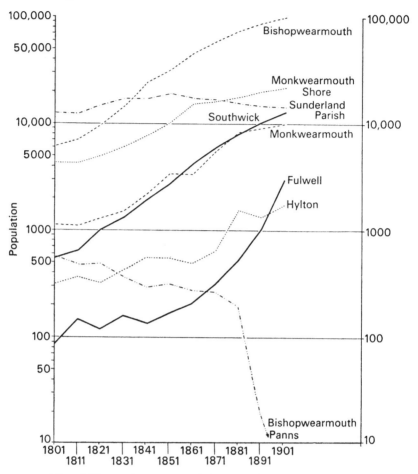

Fig. 3.6. Nineteenth-century population growth in Sunderland townships

Source : Census

comments, 'The increase of population in South Bishopwearmouth is attributed to an influx of persons from adjacent parishes, owing to the removal of a large number of houses for the construction of docks...'[1] Thus Bishopwearmouth Panns declined in population continuously to a mere 5 persons by 1901 and Sunderland Parish reached its maximum population in 1851, but thereafter showed an absolute and continuous decline. Similar to these two townships is Monkwearmouth Shore which also contained large areas adjacent to the river banks, but in addition included other less central areas. It showed a rapid growth in the first part of the century, but reached a

[1] *Census 1861.* England and Wales, I, *Population Tables,* p. 668.

saturation point after 1861 as growth occurred in areas further removed from the main industrial core of the town.

Unlike this first group of inner townships, with their falling or stagnant populations, the second group of outer townships, which includes Bishopwearmouth, Monkwearmouth and Southwick, all showed continuous expansion throughout the period. The pattern for all three is very similar: absolute expansion of population but at decreasing rates (as the semilogarithmic scale makes clear). Bishopwearmouth is an unsatisfactory areal unit in that it covered both inner (and presumably declining) areas, as well as outer areas, but the constant expansion in this and the other two townships illustrates the way in which the bulk of the population expansion was occurring in the outer parts of the town. That the rate is decelerating is largely a reflection of the general slowing-down of the growth of the whole town.

The third type of area comprises Fulwell and Hylton, both of which were peripheral and can be regarded as the third of three roughly concentric rings focused on the town centre. Here expansion of population alternated with contraction to give a much less regular profile of growth. This might be expected since the units have smaller populations than those already considered, but it is mainly due to their peripheral location. The first ripples of the town's expansion were just being felt in the early part of the century. Fulwell, lying rather closer to the centre, showed real and continuous expansion from 1841 with very considerable rates of growth thereafter. Hylton showed the same early irregular expansion but it was only much later, in the following century, that the full effects of outward expansion were felt and produced a pattern of continuous population growth.

These nineteenth-century data therefore give a clear picture of decline at the centre and expansion at the periphery, much as in the ideal single-centred town of the Chicago model. The various areas of the town appear to undergo a fairly regular sequence of growth, stability and absolute decline as the wave of the town's expansion throws its crest outwards in a widening circle. For the twentieth century, the available ward data show a more complicated pattern. That this is so is the result both of the slowing down of the total population growth and also of the influence of local-authority housing schemes. Maps for the three periods 1911–21, 1921–31 and 1951–61 show the percentage population changes in the various wards of the town (Figs. 3.7, 3.8 and 3.9). The patterns which they reveal should be considered in the context of the maps showing the age of buildings and of local-authority developments.[1] The simple outward expansion of the nineteenth century is obviously replaced by a much more complex pattern. The changes between 1911 and 1921, in particular, do not show the expected peripheral growth.

[1] See Figs. 3.5 and 3.10.

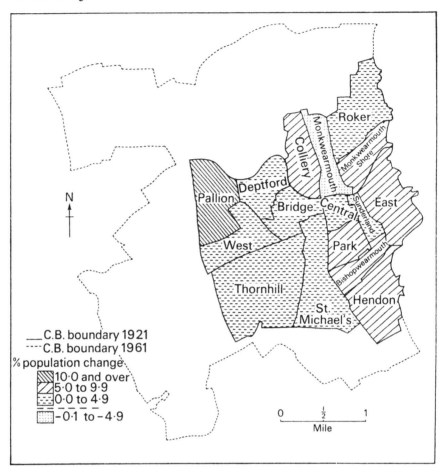

Fig. 3.7. Population change, 1911–21
Source : Census

This was a period not only of wartime, but also of general housing shortage
and high construction costs, when the widespread paucity of housing at cheap
rents gave rise to the circulars on working-class housing issued by the Local
Government Board in 1917 and 1918, and eventually to the Housing Act of
1919 which provided Exchequer subsidies for local-authority housing and
the first growth of extensive council housing for working-class families. The
imbalance in the housing market helps to elucidate the pattern of population
change between 1911 and 1921. The greatest increases, albeit small, occurred
in the inner working-class wards of the town and also in the peripheral ward
of Pallion. The inner areas in particular show no new building during this

94

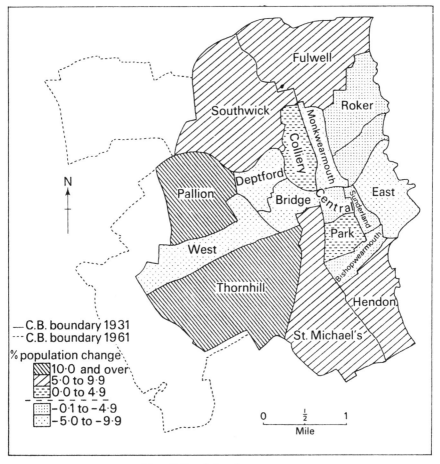

Fig. 3.8. Population change, 1921–31
Source : Census

period and indeed the increase of population was accommodated by increasing subdivision of the houses of these inner areas. It was during this
period that most of the subdivision, which is commented on in more detail
later in this chapter,[1] took place as a response to the contraction of the
housing market. The pattern therefore partly reversed the previous tendency
for contraction to occur in the inner areas except for Monkwearmouth which
lost population.[2]

[1] See below, pp. 121–3.
[2] Lancaster appears to have had a very similar housing history with a central piling-up of population in the early part of the twentieth century. See J. B. Cullingworth, *Housing in transition*
(London, 1963), ch. 2.

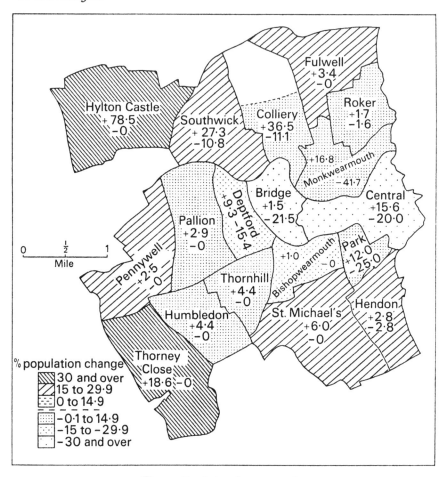

Fig. 3.9. Population change, 1951–61

Figures within each ward show the number of new (+) and of demolished (–) households in 1966 expressed as percentages of the number of households in 1961.

The northern part of Colliery Ward is left blank since it contains no housing.

Sources : Census, and Sunderland Planning Office

Whereas some of the earliest local-authority housing schemes had been of large barrack-like blocks in the old inner decayed areas, later council housing commandeered the cheaper empty space on the periphery of the town and thus the later patterns of growth again assume a more centrifugal pattern. The areas of greatest population expansion move gradually outwards from the centre of the town. Between 1921 and 1931 the greatest increases were in Pallion and Thornhill wards: between 1951 and 1961 they were in the outer areas newly incorporated into the borough. Hylton Castle burgeoned in

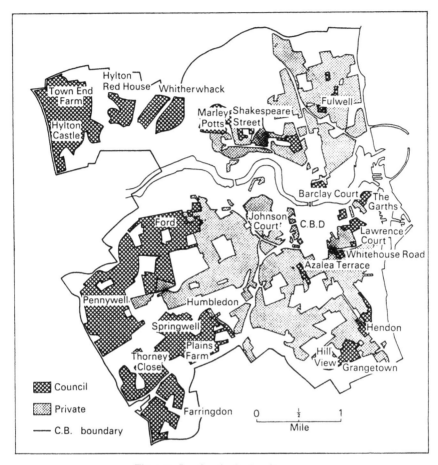

Fig. 3.10. Local authority housing areas
Source : Sunderland valuation list, 1963

population with the development of the council estates of Hylton Castle, Red House, Town End Farm and Thorney Close. Private housing likewise shows a centrifugal movement since private estates were built during the 1930s in the north of the town at Fulwell and, to a lesser extent, in the south at St Michael's, and after the war in the north at Seaburn Dene and again in the south at St Michael's.

In the 1951–61 period, therefore, even though the population of the whole town showed a small increase, eleven of the eighteen wards showed a decrease in population and all eleven were centrally located. By comparison the peripheral areas clearly show two rings of expansion: with greater rates further out and lesser rates closer to the centre.

Much of this twentieth-century expansion has evidently been due to council housing developments and, especially since council housing is a matter of some concern at a later stage, it would be profitable to examine very briefly the areas of council housing as they have developed since 1919. Seen in conjunction with the map of building ages, the map of council housing clearly shows that the progression is outwards (Fig. 3.10). The Garths, in the east end of the town, are an early development of slum replacement by high-rise flats, but they are virtually the only exception to the general rule of early developments being on the outskirts. The great majority of the building has formed an interrupted girdle partly encircling the town. Of late the pattern has changed, with redevelopment schemes now being found in inner areas such as the isolated developments in Whitehouse Court or the refurbished Azalea Terrace. Even more recent have been the high-rise flat developments such as Barclay Court, Lawrence Court and other schemes only now coming to completion in the central area. That such a partial reversal of the overall outward movement of population is likely to continue is suggested by the figures of estimated population changes between 1961 and 1966 which are shown in Figure 3.9.

The important conclusion which emerges from a consideration of population changes within the town, however, is that in general there has been a progressive outward movement which was strongest in the nineteenth century. The ebb and flow of population which is more characteristic of the present period is the result of the smaller rate of total growth and of the local authority housing schemes. This general outward movement has been focused upon the dominant central area which has developed as the Central Business District. Sunderland would therefore appear to meet the condition of the Chicago ecologists in that it has grown outwards from a single centre in spite of the fact that it began as a fusion of more than one original nucleus and, in growing, has incorporated within its expansion older centres which were once separate and independent.

PATTERNS OF LAND USE

Two main points about the town have now been put forward: first that it is essentially a town of the nineteenth century, when its growth rate was very rapid, and second that it has expanded outwards from a single dominant core. To this extent, Sunderland appears to meet two of the most important assumptions underlying the theories of classical human ecology. The towns which gave rise to the models of the Chicago ecologists shared many of the characteristics of the Victorian towns of industrial Britain. They were rapidly expanding settlements in which internal political and social organization was

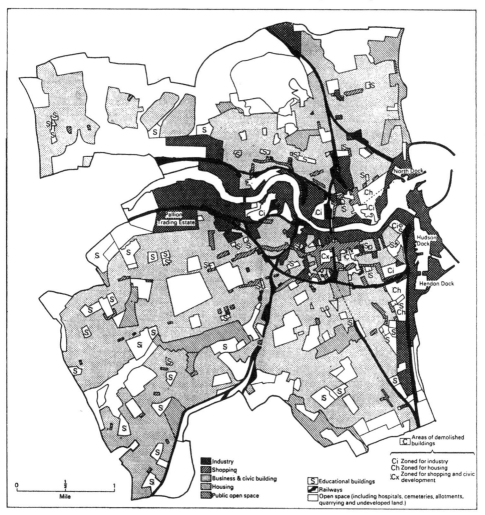

Fig. 3.11. Land use in Sunderland, 1963
Source : Sunderland Planning Office

just barely evolving and in which the free market forces of supply and demand
were able to create conditions approximating the ecological balance of plant
and animal communities. The Chicago models might therefore help to
elucidate the nineteenth- if not the twentieth-century pattern of British
urban areas. Since so great a part of the present-day town of Sunderland
carries over the physical and, to a lesser extent, the social attributes of
Victorian industrialization, such models might be expected to be particularly
applicable in analysing the town. In the remainder of this chapter it is

intended to examine the evolution of the land-use patterns of the town with a view to applying these models to the evolution of residential patterns. First, however, the general features of present-day land use must be examined.

The map of present-day land use in Sunderland shows clearly the effects of functional segregation and the sorting of more or less functionally segregated areas into a coherent pattern (Fig. 3.11). Access to the river or sea frontage has played a large role in the patterns assumed by industrial functions as one might expect since the great bulk of the heavy industry is export-orientated. The detailed siting, however, has owed a great deal to the geological and topographical features of the town. Beneath the masking layer of glacial deposits lie a range of Permian rocks whose limestone beds have produced a series of knoll-like hills in the west of the town, and towards which the highly rated housing of the nineteenth century was built. More important in the development of the industrial pattern has been the glacial diversion of the River Wear which, once a tributary of the Tyne to the north, now flows through an impressive gorge cut through the Permian scarp. At its Sunderland mouth the Wear therefore cuts a deep channel set within cliffs which rise to heights of 60 feet and more. Only at the mouth itself are there any relatively extensive areas of flattish ground adjacent to the river. This combination of geology and topography has had two effects on the industrial development of the town, which clearly differentiate it from nearby Tyneside. First, lying on the concealed coalfield, there was no development of early small pits, so that unlike Tyneside there was not the intermixture of collieries and miners' rows with other industrial works. Indeed, while Sunderland's trade depended so markedly on coal exporting, there has been only one actual mine within its boundaries. Secondly, the physiography of the river itself has prevented extensive industrial development away from the river banks. Whereas along the Tyne industry has developed along both banks to a considerable distance inland, on the Wear, space has been at a premium. While the deep incision of the river proved ideal for gravity loading of coal and the tight bends have been able to be utilized for launchings, the lack of space along the banks has had to be met by extensive construction of quays and docking facilities out into the sea at the river's mouth.

Three main industrial areas may be isolated: (i) that alongside both banks of the river; (ii) that along the eastern seaboard to the south of the river mouth; (iii) that in the west of the town, especially at Pallion. The first two areas are essentially of heavy industry; the river bank zone being an older established area than the seaboard zone. The riverside sites of the old parish of Sunderland and of Monkwearmouth at the mouth of the river, and of Southwick further upstream and Ayre's Quay at the bend of the Wear, were the earliest industrial areas and are still thronged with shipbuilding yards and coal staiths, while intermingled with these are iron-smelting industries and a host of engineering and ancillary industries particularly associated with shipbuilding. By comparison, the large complex of docks, with their associated staiths and warehouses at the mouth of the river and running south along the coast, were built largely in the second half of the nineteenth century. The North Dock was opened in 1837, Hudson Dock in 1850, the Hudson Dock Extension in 1855 and Hendon Dock in 1867. The progress over time was downstream and south along the sea coast. With the increasing size of ships and the difficulties of launching further upstream, this progression towards the less congested areas of the river is being consolidated.

The third area, in the west, is much more recent in origin. The Pallion Trading Estate has attracted to itself most of the new light industry of the town, but more recently Southwick, on the opposite bank of the river, has been developed as a second main site for such industry. With central redevelopment proceeding apace, older more central areas have now also been zoned for industrial development, as the land-use map reveals. In virtually all cases, however, the new and the old industrial areas are contiguous and form a definable industrial area: all are located with reference to the lifeline of the river in spite of the dependence of the new industries on different forms of transport. The convergence of the railways on the riverside coaling staiths and the shipbuilding facilities of the river itself has perpetuated the pattern of the nineteenth century.

Commercial and retail functions are segregated equally clearly. The Central Business District clearly provides a dominant core of specialist shops and a first-order retail centre, with second- and third-order centres being strung along the main traffic arteries and often forming nodes at pre-existing urban nuclei such as at Fulwell in the north or Southwick to the north of the bend of the Wear. Offices and business firms are almost exclusively located in an area in and around streets to the east of and parallel to the principal shopping street which lies to the south of the Sunderland Bridge. The importance of this bridge in determining the development of the Central Business District can be seen from the historical movements of the central area. Before the construction of the Wearmouth Bridge, precursor of the present bridge, in 1796, the two nuclei of Sunderland to the east and Bishopwearmouth to the west were townships linked by an east–west road along which were concentrated the retail functions of the two settlements. The building of the bridge, and the increasing hinterland which was opened, caused the fusion of the two settlements and the development of a new central business area between the two which was re-orientated in a north–

south direction.[1] Therefore throughout the nineteenth century, Fawcett Street, running south from the bridge, developed at the expense of the older High Street which ran east–west. Today Fawcett Street and its neighbouring areas have assumed the role of dominant shopping and commercial node.

RESIDENTIAL PATTERNS

Functional segregation of land uses of the sort which are shown in Fig. 3.11 is a response to the economic advantages of agglomeration and spatial proximity to locational factors such as transport. In the case of industrial and commercial enterprises, the most important locating factors are accessibility to the point of least-cost assembly of materials, the existence of certain essential site requirements, access to market potential and the external economies accruing from the proximity of functionally related activities. The patterns which are assumed are, to greater or less extent, the product of what Ratcliffe calls 'the dynamics of efficiency' within the urban area.[2] Indeed, following Ratcliffe's line of thought, and thinking of the urban area both in terms of its internal interchange of men and materials and of its external relations with its sphere of influence, the whole complex can be conceived of as an open system in which change in any one part causes a re-sorting of the functional pattern and a reassessment of the spatial relations within and without the town. Maximum entropy might be considered to be the stage at which the patterning of land uses perfectly reflected the land users' ability to pay the different rentals at different sites and at which the frictions caused to the system by the internal movements of men and materials had been reduced to a minimum.[3] Such a state is difficult to conceive of in practice, first because non-material forces can work against economic forces in towns—the classic example being Boston's Beacon Hill and second because the inputs into the system constantly change the goal which is sought. Technical changes in transport, the increase of car ownership, growth or decline of population, changes in the distribution of income, complications surrounding the transfer of land or property: all such factors affect the balance of forces which have to be considered, so that the system

[1] For similar re-orientations caused by changing patterns of internal circulation see A. E. Smailes, *The geography of towns* (London, 1953), pp. 124–5, and H. Carter, 'Aberystwyth: the modern development of a medieval castle town in Wales', *Trans. Inst. Br. Geogr.* xxv (1958), 239–53.

[2] R. U. Ratcliffe, 'The dynamics of efficiency in the locational distribution of urban activities', in R. M. Fisher (ed.), *The metropolis in modern life* (New York, 1955), pp. 125–48.

[3] Attempts to apply general systems theory to urban areas are found in L. K. Franks, 'Models for the study of community organization', *Community Development Review*, ix (1958), 1–26, and in R. L. Meier, *A communication theory of urban growth* (Harvard, 1962), esp. ch. 7. For the basic concepts see L. von Bertalanffy, 'An outline of general system theory', *Br. J. Phil. Sci.* i (1950), 134–65.

strives, without complete success, to attain a state of dynamic equilibrium. The patterns which exist are therefore partly the result of historical legacy and partly the result of contemporary forces acting imperfectly upon the structure of the land uses.

When the focus of interest is changed to types of residential area, rather than broad patterns of land use, the problems become somewhat more complex. With the important exception of accessibility to work-place, the type of area in which people decide to live is much more a product of their sets of values and of their social desires to be in proximity to people of a particular kind. The early ecologists, as has been shown, assumed a given set of value systems associated with the culture complex which they studied. It has been argued above that Sunderland in some degree shows parallels with certain aspects of the American model, and it might therefore be expected that the use of the Chicago models should be effective in analysing aspects of the evolution and patterning of residential areas within the town.

Analysis of residential segregation is bedevilled by lack of available data. One is forced to use indexes which approximate to the variable under consideration—in this case, social class. Quantification of social data and of such intangibles as 'life style' for small areas within towns is difficult even with the detailed areal data made available by the 1961 census, but for periods before this it is subject to wide margins of error. American studies have often used land values to study the segregation of urban functions. Evidence of land values is less easily come by in Britain. A department of the Inland Revenue has information, but does not make it available. Data could be collected from estate agents, but they would suffer from lack of comparability in time and incompleteness in coverage. In the absence of comprehensive data, English studies of residential segregation have largely made use of rateable values as an index of social differentials.[1] The analysis of residential patterns in Sunderland is based on gross rateable values and other data drawn from Valuation Books at three distinct dates: 1850, 1892 and 1963.[2] The objective is to analyse the evolution of residential areas, and the selection of these three dates covers, in 1850, a time immediately at the start of the town's greatest expansion, in 1892, a time at which the Victorian development had virtually ceased and the built-up area was almost as extensive as the present private housing sector and, in 1963, the recent picture. The actual dates selected were partly predetermined by the fact that complete coverage of the town

---

[1] Amongst many others see R. Glass, *The social background of a plan* (London, 1948); J. R. G. Jennings, 'A note on source material for urban geography', *East Midland Geographer*, III (1963), 212–15; E. Jones, *A social geography of Belfast* (London, 1960); R. Jones, 'Segregation in urban residential districts', in K. Norborg (ed.), *Proceedings of the I.G.U. symposium in urban geography, Lund, 1960* (Lund, 1962), pp. 433–46.

[2] For discussion of the use of these sources, see Appendix A.

could only be found for a limited number of years, so that the choice was limited. In the 1963 data only the private housing sector has been considered. The council housing areas were excluded from this initial analysis because they form a peripheral ring of low status population which could not have been foreseen by the classical ecologists and cannot fit into their models.

Before examining the patterns of valuations at these three dates it will be profitable to develop two points which arise from the use of the Valuation Book data: first, the assumptions which are made in using valuations as an index of social class: and second, the validity of the data themselves.

Assumptions involved

The use of rating values as an index of social class assumes positive correlations between social class, income and the level of rating values. Yet many factors distort so simple a premise. If the association between rating values and income is first considered, it seems reasonable to assume that, since house purchase or rental forms so large a proportion of the financial outlay of the majority of households, the price of housing will be an accurate reflection of a family's income. Yet the proportion of income which is spent on housing is not constant from class to class. It was long accepted that the higher the family income the lower was the proportion spent on housing. Reid has recently questioned this by showing that in the United States, census data from 1918 to 1960 demonstrate that the rich had *higher* income: housing ratios than the poor, so that the amount of income spent on housing would seem to rise with 'normal' income.[1] In England, Collison's Oxford data support this. He quotes Lydall and Dawson who show that 'the line relating gross rent to income for the first group (salaried and self-employed) lies above the line for manual workers, and rises more steeply as income increases, especially over the range of income £200 to £1,000'. The proportion of income spent on housing by 'white collar' tenant households is roughly between one-third and one-half as much again as the proportion spent by manual-worker households, at each comparable level of income.[2]

Assuming this to be true, the effect on rating values will be to exaggerate the differentials between the hierarchy of incomes and to emphasize the distinctions between the borderline zone of the blue- and white-collar groups.

Considering the association between income and social class, the literature

[1] M. G. Reid, *Housing and Income* (Chicago, 1962). Reid distinguishes 'normal' income (i.e. the basic or expected) from 'actual' income (which may be greater or less depending on bonuses, illness, etc.).

[2] P. Collison, 'Occupation, education and housing in an English city', *Am. J. Sociol.* LXV (1959), 599–605; H. F. Lydall and R. F. F. Dawson, 'Household income, rent and rates', *Bull. Oxf. Univ. Inst. Statist.* XVI (1954), 124.

suggests a considerable overlap between the lower echelons of the white-collar workers and the higher echelons of the manual classes. This, however, will be counteracted by the tendency, noted above, for the variable proportion of income spent on housing to differentiate between these two groups.[1] The combination of these two sets of distortions would therefore tend to cancel each other out. Thus though the associations between social class, incomes and rating values are not simple, it would seem that rating values can be expected to be a useful approximate index of class differences. Indeed, for the 1963 data it is possible to quantify this association since the 1961 census provides information on socio-economic groups for enumeration districts. Calculating the median gross rateable value for each enumeration district within the town, the simple correlation coefficient between this and the average social class in each enumeration district (derived from the Registrar General's socio-economic groups) in 1961 was $r = 0.867$ for the private housing sector of the town.

In spite of this high positive association between class and rating values, we might note other reservations associated with the use of the valuations. Households include differing proportions of earners and non-earners: the large family is at a disadvantage compared with the single-person household or the childless family as regards available income per head of the total family. Again, rating values will fail to differentiate different patterns of 'life style'. Sociologists suggest that the life styles of the rooming-house area and the suburban area are of very different kind. Wirth would say that the rooming-house area is further developed along the continuum of 'urbanism'. Such areas typically have an admixture of high status single or newly married households and of lower status households, and the use of rating values will thus only give a poor reflection of the objective social class groupings of such heterogeneous areas, even though it may give leads to less tangible factors.

Bearing such reservations in mind, however, it would appear that rating values are the most discriminating *single* index of social class which is available for periods before census enumeration district data were made available in 1961. Hoyt and Schmid consider rental value the most useful index, and while others may be as discriminating (Schmid, for example, in a later article considers that educational achievement is a better index of class than is mean rent),[2] for studies in Britain prior to 1961 such variables were not readily at one's disposal.

---

[1] The Duncans show that, in Chicago, clerical workers, with a lower income than foremen-craftsmen, nevertheless live in higher-priced housing. Cf. O. D. and B. Duncan, 'Residential distribution and occupational stratification', *Am. J. Sociol.* LX (1955), 483–503.

[2] Hoyt, *The Structure and growth of residential neighborhoods, op. cit.*; C. F. Schmid, 'Generalizations concerning the ecology of the American city', *Am. Sociol. Rev.* XV (1950), 264–81; and C. F. Schmid, E. H. MacCannell, and M. D. van Arsdol, Jr., 'The ecology of the American city: further comparison and validation of generalizations', *Am. Sociol. Rev.*, XXIII (1958), 392–401.

Urban analysis

Validity of the data

Gross rateable value data for 1850, 1892 and 1963 were drawn from Valuation Books at the three dates. The 1963 data are undoubtedly the most reliable, being a re-valuation based on 1962 levels of housing. The 1850 data must be interpreted with the greatest caution since little uniformity guided the valuers at this time. The principles of rating have grown on a very *ad hoc* basis. Their development has been slow and spasmodic and in the main has arisen from countless judicial decisions prompted by practical problems as they presented themselves. The Poor Law Commissioners in their first Report of 1834 drew attention to the lack of uniformity prevailing up and down the country in the making of poor rate assessments, on which the local rates were based. The usual practice was to take the annual value of property as the basis for the rating assessment, but each parish was a law unto itself in this matter. The 1836 Parochial Assessment Act was passed to attempt to secure some degree of uniformity and made the annual value the legal basis for assessment. The 1850 data are covered by this Act, but even so there is no certainty that the returns of all parishes are truly comparable. It seems, in fact, from a study of the three parishes involved at this date, that one of them used a lower rate of assessment than the other two. The figures have therefore to be interpreted cautiously.

The 1892 data are covered by the 1862 Union Assessment Act, which was passed specifically to give a uniform and correct valuation of parishes in the Poor Law Unions of England. The basis of assessment was not changed by this Act: it remained the annual value of property, but a gross value was introduced from which the rateable value, as defined by the Act of 1836, was calculated. This gross value in the 1892 returns meets our need for a figure which is comparable from area to area.

Many doubts have been expressed recently as to the validity of the present-day rating assessments. As is noted in Appendix A, the Milner Holland Report pointed to discrepancies between rating assessments and the actual level of rents charged in the London area. One might expect such discrepancies to be greater in an area such as London where the greater housing shortage and the large demand for furnished lettings combine to raise the market value of such types of property. In Sunderland the position is less acute and one might imagine that such a criticism of rating values would be less valid. In any case, the present use of valuations is to measure social class rather than actual rentals and, as the high correlation noted above would suggest, the general association would seem well established. Since the 1963 valuations are an assessment based on 1962 conditions, the old criticism of valuations before the re-assessment no longer applies. The great advantage

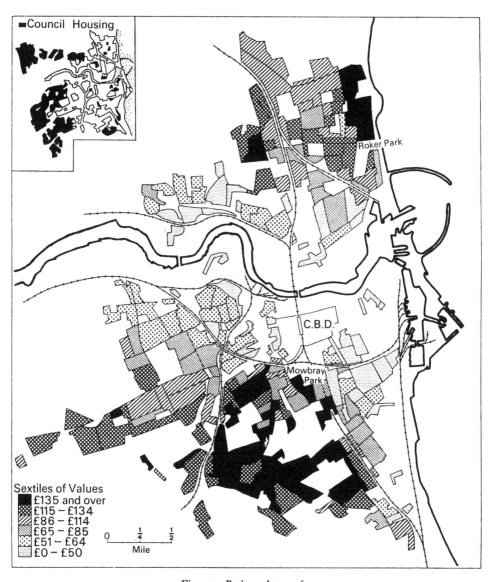

Council Housing

Roker Park

C.B.D.

Mowbray Park

Sextiles of Values

£135 and over
£115 – £134
£86 – £114
£65 – £85
£51 – £64
£0 – £50

0   ¼   ½
     Mile

Fig. 3.12. Rating values, 1963

Only the private housing areas are considered. The additional areas of council housing are shown in the inset.

Source : Sunderland valuation list, 1963

of using the Valuation Lists is that their compilation is a central and not a local government function and we can therefore expect standardization. Valuation Officers calculate the gross rateable value of properties in terms of 'the rent at which the hereditament might reasonably be expected to let from year to year, if the tenant undertook to pay all usual tenant's rates and taxes...and if the landlord undertook to bear the cost of the repairs and insurance, and other expenses, if any, necessary to maintain the hereditament in a state to command that rent'. In making this assessment, two main conditions are considered: first the physical nature and state of the building; second, the advantages accruing to a particular location and site.

PATTERNS OF RATING VALUES

The 1963 Valuation Lists were made available by the Sunderland Rating Office. Private housing areas and council housing areas are entered in separate sets of volumes, which proved useful in making a clear distinction between the two sectors. From the information provided two sets of data were extracted: the median gross value per street (or part of a street where wide intra-street variation was found); and the proportion of dwellings per street (or part of a street) which were recorded as 'houses', as distinct from 'flats' or 'rooms'.[1]

The median gross value was the main focus of interest. This was plotted in detail to a scale of 1:19,560, and on the basis of this map the town was divided into a large number of small areas delineated on the basis of maximizing homogeneity with respect to the valuations. In all, 274 areas of varying size were isolated; 227 of which were of private housing, and 47 of council housing. In the following analysis, the council house sector has been ignored for reasons which are elaborated in chapter 4. For each of the private housing areas an average value was then calculated and these figures form the basis of the analysis of rating values. These figures are shown diagrammatically in Fig. 3.12.

The second measure—proportion of 'houses'—was calculated as a rough measure of the degree of subdivision of dwellings. The Valuation Book distinction between 'houses' and 'rooms' or 'flats' is a useful index of the conversion of dwellings into multi-occupied habitations, and thus provides a measure of the 'invasion' of areas by people of successively lower economic and social status. This process of 'invasion' played a large part in the thinking of the early writers of the Chicago school, and the degree of housing subdivision traces this process by isolating those areas which have undergone sequential change of population from wealthier to poorer people. Fig. 3.13 plots the distribution of this characteristic for 1963.

[1] See Appendix A for a discussion of these terms.

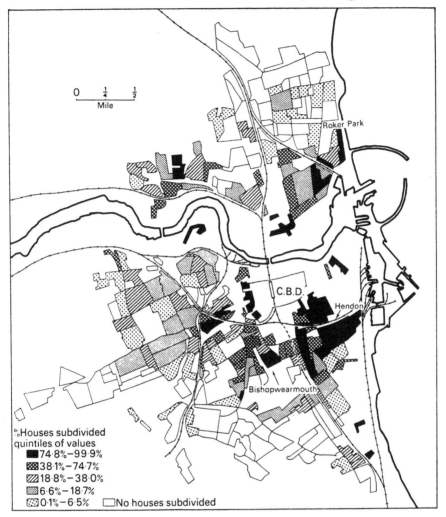

0 ¼ ½
Mile

Roker Park

C.B.D.

Hendon

Bishopwearmouth

%Houses subdivided
quintiles of values
■ 74·8%−99·9%
▨ 38·1%−74·7%
▨ 18·8%−38·0%
▧ 6·6%−18·7%
▦ 0·1%−6·5% ☐ No houses subdivided

Fig. 3.13. Subdivision of Housing, 1963
Source : Sunderland valuation list, 1963

Taken in conjunction, the two maps illustrate the partial relevance of the two principal ecological models—the concentric-circle model of Burgess and the sectoral model of Hoyt. The map of rating values shows a clear distinction between the patterns developed to the north and the south of the river. To the north, the Burgess concentric-circle model seems most appropriate. Close to the industrial zone of the river is a zone of low values which increase gradually northwards up to the railway branch line running to the North Dock. To the north of this railway, values are noticeably higher and form a highly rated area which extends to the northern boundary of the town. At only one point does this highly rated area bend southwards towards the Central Business District. This is the section which follows the main north–south road into the centre of the town. Two interesting exceptions to this concentric pattern may be noted. First is the small area in Fulwell in the north which forms a nucleus of slightly less highly rated houses. This marks the centre of the old settlement of Fulwell which has been incorporated into the town in its outward expansion. The nucleus now plays the role of secondary shopping centre as Fig. 3.11 shows. Second is the area fringing the eastern seaboard which has values consistently higher than those inland. The pattern of values is here tending to form a distinct sector running adjacent to the sea, rather than conforming to the concentric zone ideal.

To the south of the river, by comparison, the Hoyt pattern of sector wedges seems more appropriate. From east to west there are four clearly differentiated sectors running outwards from the centre of the town. In the extreme east is a sector which fringes the coastal industrial belt and which has low values. It reaches out from the artisan area of the old parish of Sunderland southwards to the southern boundary. This first sector ends fairly abruptly along the line of the main road running south from the town, and to the west of this road is a highly rated sector with, at its heart, the large open spaces of a farm, a public park and the grounds of a cricket club. To the west, this sector merges irregularly into a sector of medium-high values which runs westwards from the centre of the town and to the south of the east–west road leading towards Durham. The final sector is to the north of this road in the westward-running area between this road and the industrial zone flanking the river.

Superimposed on this pattern is the distribution of subdivided housing which forms a pattern common both to the areas north and south of the river. The zone of highly subdivided houses forms a concentric belt fringing the Central Business District and the riverside industrial area. In the north there is a narrow strip of highly subdivided houses running to the north of Monk-wearmouth Shore, and a small area at the townward apex of the highly rated area on the seaboard. To the south of the river, the zone of subdivided housing is much more extensive and forms a broad girdle encircling the Central Business District, narrower in the west, but of considerable width in the east. Thus the pattern of outward-spreading decay which Burgess stressed would appear at first sight to be vindicated.

The analysis of these patterns, however, can be made clearer if the historical evolution of the various sub-areas within the town is traced in detail. Fig. 3.14 shows the pattern of rating values in 1850 for the three parishes of Bishopwearmouth, Monkwearmouth and Sunderland. Segregation of residential areas is clearly apparent. The three nuclei of which the present town is principally composed had coalesced by this date around a Central Business District which was not clearly defined, but which centred around the east–west High Street which is shown on the map as the principal shopping street. The most interesting feature of the map is the clear separation of the highly rated area of 1850 in an area which is now the southern part of the Central Business District and which today contains most of the office and commercial functions of the town. The low rated area is found in the east end

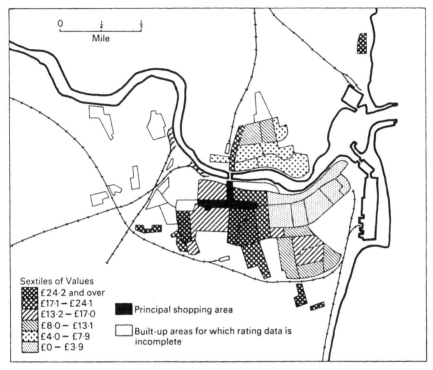

Fig. 3.14. Rating values, 1850

Source : Assessment for the relief of the poor, Sunderland, 1850

of the town in the parish of Sunderland. The pattern which emerges is rather confused. The fact that the Central Business District appears not to have welded the town into a single, definable functional unit is seen in the distinct patterns which have formed round each of the nuclei, in a way reminiscent of the multiple-nuclei theory of Harris and Ullman.[1] There are, however, elements of both the Burgess and Hoyt models within this. The highly rated area of Bishopwearmouth, for example, forms a sector reaching out from the

[1] C. D. Harris and E. L. Ullman, 'The nature of cities', *Ann. Am. Acad. Polit. Social Sci.* CCXLII (1945), 7–17.

main shopping area towards the open ground to the south. Another highly rated area in the north forms a partial sector following the main road leading north towards Newcastle. The isolated area of highly rated housing in the west of the town marks the core of the old village of Bishopwearmouth with highly rated houses built round the old village green. At a later period, this was to become a low rated area as the industrial areas expanded around the coal staiths of the river and poor housing was built in the old village nucleus. In the east of the town, a rather different pattern obtains with elements of the concentric rings apparent. The low rated area of Sunderland parish is succeeded to the south by a belt of mixed valuations which includes large areas of quite highly rated housing. This tends to increase in value outwards from the river, so that to the south of the east–west railway line there are two areas with average values within the highest sextile of values. This mixed area of medium-high ratings forms Upper Hendon (to the north of the railway) and Lower Hendon (to the south of the railway), whose subsequent changes in valuation are discussed in detail below. In 1850 the area was part of a concentric pattern in this eastern part of the town.

The 1892 values, plotted in Fig. 3.15, show how these sectors have been developed and further differentiated by the end of the maximum period of growth of the town. To the south of the river there are five conspicuous features of change. First, the Central Business District has been greatly expanded and consolidated. Few of the houses in the inner highly rated area of 1850 remain. Most of the area has become clearly differentiated as a business and retail centre, which forms a core around which the town is orientated and by which it is welded together. This is largely the outcome of the population growth of the town itself. It would appear that below a certain threshold size, which is not clear in terms of absolute numbers, a clearly defined Central Business District does not appear. By 1892, with a population of 131,015, such an area has appeared and it is interesting to note that it has been in the area of the 1850 highly rated sector that this development has occurred—very much in accordance with the ideas of Hoyt, who suggested that the point of origin of the highly rated sector is at the commercial and business heart of the town, so that it is into this sector that business expansion occurs. A good example of this invasion of business functions into once-residential highly rated areas is seen in the admixture of highly rated houses and offices within this area. From the Valuation Book for Central Bishopwearmouth in 1892 we can reproduce a section of John Street, which lies immediately to the east of the present-day main shopping axis in Fawcett Street. In 1850, John Street was wholly composed of private houses with a median value of £36 and with individual houses valued as high as £60 and £65 and was, in fact, in an area falling within the highest sextile of rating values (see Fig. 3.14). A section of the 1892 returns, however, reads as follows:

Description of Property	Number	Gross Value (£)
Lecture Hall	30	40
House	30	63
House and Stable	31	80
Office and Stable	32	80
Office	32	16
County Court		
House	45	60
House	46	50
Coach House and Stable	Back 46	10
Office	47	20
Office	47	75
Office	47	15
Office	47	43
Office	47	25
Rooms	47	10

A few highly rated purely residential houses remain, but many have been subdivided and invaded by encroaching offices, public buildings and sub-divided rooms. By 1963, the street had been completely submerged by such office functions and today it forms the core of the business area, including such buildings as the local authority Education Offices and a large number of insurance firms.

The second noteworthy change during the period is the way in which the highly rated area of 1850 has continued its original line of development reaching out, by 1892, towards the higher ground in the south of the town. Third, the small nucleus of highly rated houses in Bishopwearmouth has been engulfed within a low rated area stretching westwards along the line of the river industries. Fourth, a rather similar 'swamping' has occurred in the case of the 1850 mixed medium rated areas in Hendon to the south of Sunderland parish. Here the low class housing of the east end of the town appears to have formed a sector running southwards parallel to the coast. Finally, in the west of the town, the origin of the medium rated area of 1963 can be traced developing along the line of the main road running westwards towards Chester-le-Street and Durham. The ideas of Hoyt would seem to be vindicated, since he placed great emphasis on the importance of main communication arteries in determining the location of the more highly rated sectors. The importance of the road is emphasized by the distinction between the values of houses within this western area. Houses to the north of the road are valued at below £15, while those built along the line of the road to the south of it are valued at between £15 and £30, and in the case of one street, at above £30.

Before developing the analysis of these patterns revealed by the rating values, however, the nineteenth-century values themselves should be

Fig. 3.15. Rating values, 1892
Source: Valuation assessment, Sunderland, 1892

examined in more detail to validate the data themselves. The use of rating values as an index of social class has already been discussed. The high positive correlation between social class and rating values for 1963 has been shown but, since lack of direct information on social class for the earlier periods makes it impossible to test the validity of the nineteenth-century valuations by using similar correlation techniques, it will be profitable to compare the rating value patterns of the nineteenth century with evidence drawn from other sources. Three sets of data have been used; the distribution of various types of schools, the distribution of pawnshops, and the distribution of diseases. The selection of these indexes was based on the attempt to find the most discriminating and accessible social measures of class, and these three pieces of information can be shown to meet this need.

Distribution of schools

The type of school found in an area is a discriminating indicator of the social composition of the area, since young children are less likely than adults to travel far from their homes. Especially where very young children are concerned, the catchment area of each school is likely to be drawn within a tight radius of the school itself. This is likely to be particularly true for the nineteenth-century period under consideration since communications were so much less well-developed, and while this led to children in rural areas walking long distances to schools, in urban areas it doubtless led to an intimate association between home place and school place. The data on schools mapped in Fig. 3.16 were drawn from the report of the School Board of the district of Sunderland in 1873. In recording the schools which existed at this time, the Clerk of the School Board drew a distinction between two types of non-public elementary school: 'one where the fees exceed 9*d*. per week, and which may be called high status schools, and the other where the fees are under that amount'.[1] The non-public elementary schools with fees of over 9*d*. may therefore be taken as indicators of high status areas. The non-public elementary schools with fees below this, together with the public elementary and charity schools, will be taken as indicators of low class areas.

The distribution of high status schools corresponds well with the distribution of the highest two sextiles of the 1892 rating values. Both form a triangular wedge running southwards from the centre of the town. The intermingling of high and low status schools on the eastern border of this wedge illustrates well the process of invasion of which more is said below. Outside this principal highly rated area, the remaining high status schools by no means contradict the pattern of rating values. Four schools are found to the north of the river in an area which is not consistently highly rated. However, they are all found along the main road leading north towards Newcastle which includes a small area of highly rated houses and, in the absence of information on the catchment areas of these schools, one can hazard the guess that they served the highly rated areas which lie to the north of the more densely built-up area of the town, in an area where the number of houses was insufficient to justify local schools. In terms of overall accessibility, these schools represent the best location to serve these widely scattered peripheral residential areas. To the south of the river, there are two high

[1] *The school board of the district of Sunderland, clerk's report, 1873,* Sunderland Public Library, p. 4.

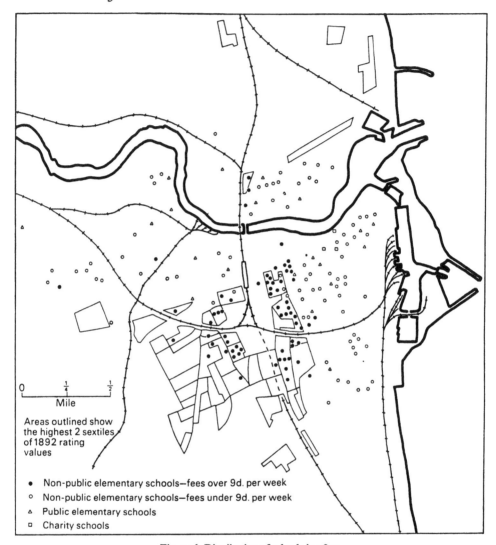

Fig. 3.16. Distribution of schools in 1873
Source: The school board of the district of Sunderland, clerk's report, 1873

status schools in the west of the town serving the area of medium-high rating values which has been noted running out to the west. Secondly, there are a number of schools to the east of the main highly rated wedge which fall outside the area of high valuations. To a large extent the apparent discrepancy here is accounted for by the difference of date of the school data and the valuation data. There appears to have been a westward shift of the highly rated area so that in the 1870s these marginal areas which contained high status schools in 1873 would have been more highly rated at that time. Only one of the high status schools cannot be reconciled with the evaluation data. This is the school in Sunderland Street in the northern part of Bishopwearmouth, lying just to the south of the river. Sunderland Street, in 1892, had a median rating value of only £7 and the adjacent areas are composed of streets none of which had values of over £10. While, as the map of 1850 values shows, this area was then highly rated, we cannot think of the school as a high status one now in decline (as applied to those schools on the eastern margin of the highly rated area) since the school rolls show an increase in attendance between the years 1871 and 1873. Nor is its location particularly accessible to other high class areas since all the adjacent parts have suffered a high degree of invasion and decay.

With this one exception, therefore, the distribution of the high status schools confirms the pattern of rating values. The non-public schools with low fees, the public elementary schools and the charity schools, with only four (explicable) exceptions, are found in low rated areas, and the public and charity schools are found, without exception, in the very low rated areas of the eastern and central parts of the town.

Distribution of pawnshops

Information on the distribution of pawnshops was drawn from town directories for dates contemporaneous with the 1892 valuations. Pawnshops are again a discriminating index of local social conditions since they are unspecialized, local facilities with a high frequency of customer usage.[1] The population which they serve can therefore be expected to be an essentially local one and the presence of pawnshops would tend to indicate low social class in an area. Furthermore, since the initial investment costs of such small establishments are low, they tend to be established and wound-up the more readily as local social conditions change. Therefore they are more diagnostic of local social circumstances than if they were larger and more permanent establishments. The distributions for 1883 and 1899 thus show great changes in detail, yet the overall pattern is remarkably constant (Fig. 3.17). In 1883 there were 53 pawnshops in the town, in 1899 there were 59. In no case is any of these found within the area marked by the highest two sextiles of rating values, although the geographical spread is fairly wide throughout the remainder of the town. Again, one of the interesting features of the distribution is the way in which, at the later date, pawnshops can be seen to be

[1] Pawnshops are used as an index of social conditions, for example, in Glass' study of Middlesbrough (Glass, *The social background of a plan*).

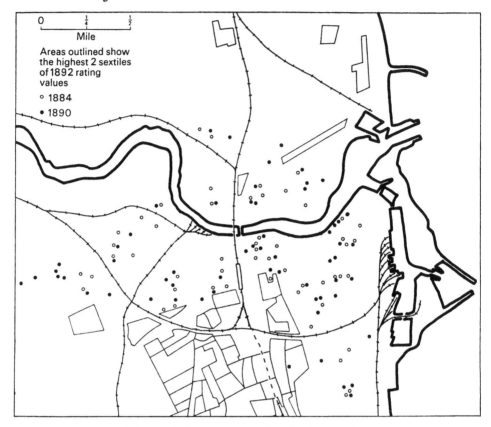

Fig. 3.17. Distribution of pawnshops in 1884 and 1890
Source: Ward's Directories of Sunderland, 1884 and 1889–90 (London, 1885 and 1890)

encroaching on the eastern margins of the highly rated wedge to the south
of the town in Hendon. This provides additional evidence of the progressive
invasion of this area by people of lower social class causing falling status and
increasing decay.

Distribution of diseases

Finally, the distribution of diseases adds further confirmation to the nineteenth-century rating valuation patterns. The information is fragmentary,
but in all cases it substantiates the patterns of nineteenth-century rating
values (Figs. 3.14 and 3.15). The outbreak of cholera in 1866 (Fig. 3.18)
emphasizes the unhealthiness and insanitary state of the crowded east end
of the town, which the 1850 values showed to be by far the lowest rated area.
No deaths occurred in highly rated areas, and in fact the number of deaths

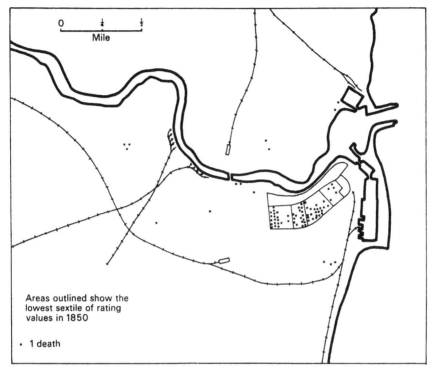

Areas outlined show the
lowest sextile of rating
values in 1850

· 1 death

Fig. 3.18. Distribution of deaths from cholera, July–December 1866
Source : Handwritten report in the care of Sunderland Rating Office

outside the eastern parish was too few to allow more than the flimsiest conclusions to be drawn for these areas.

The map of infectious diseases in 1890 (Fig. 3.19) likewise shows the highest rates to be in the eastern part of the town. In showing the distribution, parish populations for 1891 have been used to provide the most detailed breakdown of the population. The fact that there is not a perfect association between the incidence of disease and the valuations is not entirely unexpected. The diseases are all infectious and, because of this, the distribution of deaths from these causes is a poor index of social differentials since it measures variables which are dependent on physical contact as well as poor living conditions. It is not sufficient that an area is poor; it must also have some contact with the disease in question before the poverty is underlined. Low social class is thus a partial prerequisite, but not necessarily a precipitating factor in the high incidence of deaths. Nevertheless, the point to be made is that in no case is a high incidence of death found in a highly rated area. The parish of Christ Church, in fact, which embraces the area of highest gross

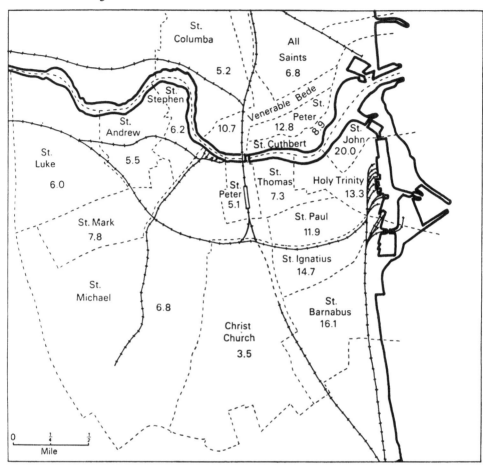

Fig. 3.19. Mortality rates from infectious diseases, 1890

Figures shown are deaths per 1,000 population in each parish. Diseases are as follows: scarlet fever, enteric diseases, typhus, measles, diphtheria and diarrhoea

Sources: Infectious Diseases in Sunderland, 1890 (being a map, in Sunderland Public Library, illustrating the report of the Medical Officer of Health, 1890): and *Census* (parish populations)

rateable values, has the lowest incidence of deaths. Thus again the evidence would seem to substantiate the rating patterns.

Having added validity to the patterns of nineteenth-century valuation, these patterns can now be examined in more detail.

Movement of the highly rated areas

Hoyt, in his analysis of residential areas, lays great stress on the location and movement of the highly rated sector which, he argues, determines the pattern of residential areas for the whole of a town. The development of the highly rated area in Sunderland illustrates the force of much of his argument. In 1850, the highly rated sector had developed at a point close to the commercial and business area of the Central Business District, as indeed Hoyt would predict, and from this point of origin it extended its line of growth consistently outwards towards the high and open ground to the south. This line of expansion is continued through 1892 until, by 1963, it forms a distinct sector to the south of the town some distance from the Central Business District.

In the north of the town, the relatively highly rated area developed at some later date. The origin of the broad concentric zone of high values is seen as early as 1850 in isolated houses of high value, but to some extent such isolated dwellings were evenly scattered in all the quarters of the town. Here again is a conflict between the centrifugal forces stressed by Burgess and the directional forces stressed by Hoyt. Before the clear segregation of distinct residential patterns, the peripheral areas tended to be occupied by a scattering of highly valued houses. In the north of the town, this concentric pattern is continued through to the present, as has been shown, whereas in the south a sectoral pattern increasingly asserts itself, and only certain parts of the old peripheral ring continue to be occupied by highly rated dwellings.

Subdivision of houses

It has been suggested above that the subdivision of houses can be used as a measure of 'invasion' of the kind stressed by the early Chicago writers. To the extent that the degree of subdivision in 1963 shows a discontinuous ring of highly subdivided houses around the central parts of the town (Fig. 3.13), the process of invasion seems to have assumed a fairly regular concentric form. Certainly there has been an outward movement from the low rated areas close to the river into areas of larger houses which were once lived in by wealthier people. The degree of subdivision in 1892 (Fig. 3.20) shows that the pattern is of long standing. The concentric ring pattern is clearly evident and it was noted earlier how the rates of population growth showed a similar outward concentric pattern.[1] The process of subdivision in 1892, however, has proceeded much further from the centre in the eastern part of the town than elsewhere and this may be a reflection of the fact that in 1850 this eastern area was composed of relatively highly rated houses so that at this time a concentric pattern was developing which was subsequently to become, in this southern half of the town, a sectoral pattern. Thus, while there has been a general outward concentric invasion of the Burgess type, it would seem that a

[1] See above, pp. 91–8 and Figs. 3.6, 3.7, 3.8 and 3.9.

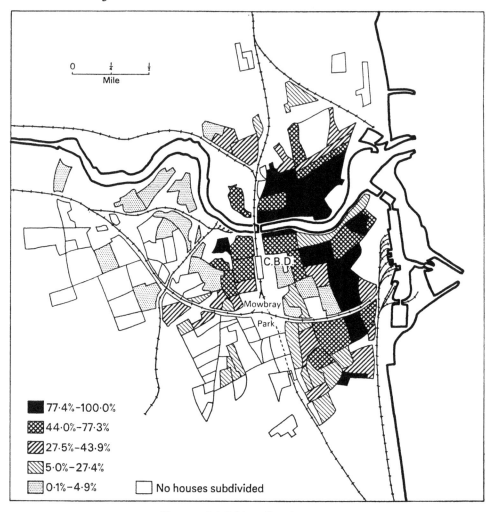

Fig. 3.20. Subdivision of housing, 1892
Source : Valuation assessment, Sunderland, 1892 ·

directional component has to be distinguished within this movement of subdivided housing. Indeed, while the subdivided housing does form a concentric ring, there are in fact two principal types of subdivided area; first the classic rooming-house area to which typically the single or young married, transient and mobile population is attracted and in which great social heterogeneity is found; and secondly, the more residentially stable type of area which contains families of manual workers and is of lower and homogeneous social status. To a large extent the existence of this second type of area in

Sunderland is a consequence of the town's housing shortage and the low incomes of much of its population.

It is with respect to this difference between two types of subdivided area that we can isolate the directional component in the evolving ecology of the zones of change and decay. There are four principal areas of highly subdivided housing. Two of these are of the first type, and two of the second type of subdivision. To the south of the river, there is a rooming-house area in Bishopwearmouth close to the centre of the town, and secondly there is an area to the east of it, in Hendon, which is of low-class subdivided housing.[1] To the north of the river, there is a small rooming area near Roker Park close to the sea, and secondly, a more extensive area of low-class subdivided housing stretching to the west adjacent to the industry along the banks of the river.

Both to the north and to the south of the river, the rooming-house areas are therefore at the townward apex of the high rated sectors, very much in accordance with Hoyt's ideas. Especially in the more extensive area to the south of the river, these areas contain the classic type of rooming-house tenant with an admixture of unsubdivided houses lived in by wealthier families or older persons and, cheek by jowl, subdivided housing lived in by young single persons or young childless couples. A high proportion of the student population of the town is found in the southern rooming-house area.

The second type of subdivided area has very different characteristics. The change of population has been much more complete so that few unsubdivided houses remain and none of the houses has a high gross value. Thus, for example, one of the sub-areas of the southern rooming-house area has a median gross value of £115 and only 27 per cent of its houses subdivided, another has a gross value of £140 and 56 per cent of its houses subdivided. By contrast, the low class subdivided area in Hendon to the east has few areas with a median gross value of over £50, and consistently over 50 per cent of the houses are subdivided. The area is predominantly of one class with an overwhelming proportion of families of low status.

Ecological changes and decay, 1892–1963

The examination of the evolution of the ecology of the town can be carried somewhat further by looking in more detail and with more precision at the changes which have occurred between 1892 and 1963. In attempting a comparison at the two dates the examination has to be restricted to a relatively small part of the town—that part which was residential at both dates. The

[1] These two types of subdivided area—the rooming-house area and the low-rated subdivided area—are examined in more detail in the field surveys reported below, where the differences between them clearly emerge. See the descriptions of Bishopwearmouth and Upper Hendon in chapter 5.

expansion of the industrial area and the development of the Central Business District have engulfed areas close to the river and those close to the central part of the town which were residential in 1892. Conversely, many of the residential areas of 1963 were open country in 1892. In the intervening zone,

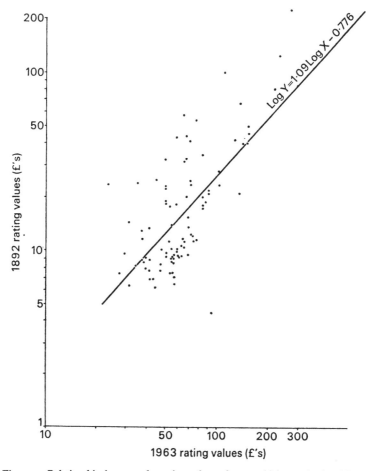

Fig. 3.21. Relationship between the rating values of areas which remained residential between 1892 and 1963

Sources: Valuation assessment, Sunderland, 1892 and 1963

the sub-areas which were delineated as homogeneous with respect to their 1963 rating values have been used to calculate the values in 1892. This produces a total of 92 areas with identical boundaries at two dates. Plotting the two values for each area, it can be seen that, while there has been a considerable increase in absolute values, there has been little *relative* change

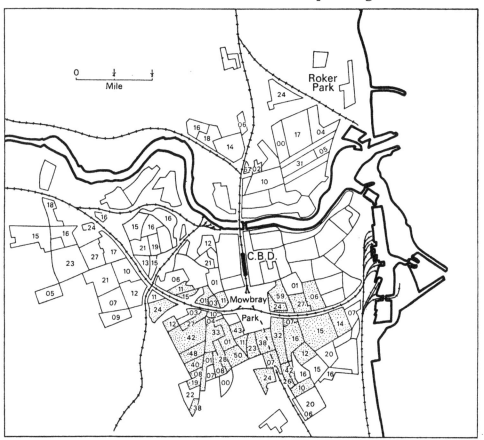

Fig. 3.22. Changes in rating values between 1892 and 1963

All the 1892 residential areas are shown in outline. Positive or negative change in rating between the two dates is indicated by decimal fractions in those areas which were residential at both dates. The figures refer to deviations from the regression of log *y* on log *x* (see Fig. 3.21). Areas showing *negative* change are stippled

between the two dates (Fig. 3.21). The correlation coefficient (using logarithms of values) between the pairs of values at the two dates is *r* = 0·687. The rate of change over all of the 92 sub-areas which were residential at both dates thus appears to be constant. Yet while most of the areas show little relative change of value, there are some which lie at some distance from the regression line of log *Y* on log *X*. Thus the change in *relative* values can be measured by each area's deviation from this regression line (which marks the line of 'expected' change in value between the two dates). Values lying below the line have 1963 valuations which are higher than the expected figure; those below the line are lower than expected. Fig. 3.22 shows the distribution of this change in cartographic form. The values entered in each area are the amounts—either positive or negative—by which the areas deviate from the regression of log Y on log X. Those areas which are shaded are ones in which the change in relative values has been negative. In other words it is in these areas that the decay and invasion of the period 1892–1963 has been concentrated. The pattern is startlingly dramatic. Decay has been limited to a sector in the south of the town reaching east almost to the sea coast: that area in fact which has been isolated as having in 1963 a large proportion of subdivided houses, and including both the rooming-house area in Bishopwearmouth and the low-class zone of decay in Hendon. To the extent that the decay is found in the rooming-house sector, the Hoyt model of urban growth is clearly vindicated. Invasion of a different succeeding population has proceeded along a well-defined path in the highly rated sector outwards from the centre of the town.

Yet this is by no means the whole picture since the area of decay extends further to the east along a concentric ring which formed the periphery of the town in 1892. It is this area which today forms the low status subdivided area of Hendon. Judging from the patterns of gross values at the three dates (Figs. 3.14, 3.15, 3.12), from the pattern of relative change (Fig. 3.22) which shows successively lower amounts of loss of relative value as one moves further west, and from the evidence which has been noted above, of the patterns of schools and pawnshops,[1] the process of decay seems not to have proceeded outwards from the centre of the town, but in a westerly direction away from the coast. The earliest and greatest decay in the twilight zone of Hendon has been in the east and has spread progressively westwards. This is less easy to reconcile with the expected pattern of the Hoyt model—or indeed with the Burgess concentric model. Explanation must be sought elsewhere.

The 1850 values show Hendon to be of reasonably high value, forming, with the low-rated area of Sunderland parish, a concentric pattern in accordance with Burgess' theory. By 1892 it can be seen to be undergoing decay to an area of low values. By 1963, it has clearly become, not part of a concentric zone, but a low-class sector conforming to the Hoyt model. Yet the decay has proceeded in a westerly direction which calls for further examination. Contemporary records of the period can help elucidate this picture of westward-spreading decay in the Hendon area. Southern Hendon in 1850 was an area of largely open country dotted with but few houses. The sea coast was something of a resort. Waters comments, 'Between 1820 and 1840, visitors occasionally came to Hendon...for bathing and to benefit from the chalybeate waters of the spa'. In addition there were in the area several country mansions which 'evinced the popularity of Hendon as a place of

---

[1] See Figs. 3.16 and 3.17, and pp. 115–18 where this westward invasion is commented upon.

residence for the wealthy'.[1] These conditions were beginning to change as early as 1850. The construction of the South Docks in 1850 had swallowed up the spa and given the neighbourhood an industrial character. The grounds of the older large houses became surrounded by small artisan houses and the railway to Seaham and the Hartlepools passed through the area. Whereas in 1850 houses in Hendon contained only one family, Waters, writing as a contemporary, notes that by 1900 'a single-family mansion is more or less a rarity...two, three, four or even five families, consisting in extreme cases of 25 or 30 individuals live in an 8-roomed house'. He comments, 'Fashion in Hendon, as elsewhere, is ever stepping westwards'.[2]

To account for this process of westward-moving decay one has to look beyond Hoyt's analytical model and consider the role of industrial development in moulding the social values which are given to areas. While he stresses the importance of the movement of the highly rated sector, this overlooks the fact that the distribution of industry is one of the prime factors in moulding the ecological structure of such industrial towns as Sunderland. As Hawley argues, 'Since familial functions constitute a relatively unintensive use of land, they cannot compete successfully in most instances with business and industrial functions for the most accessible locations'.[3] It was this factor of the location of industry which was so forcibly stressed by Davie as being the most important in determining urban residential patterns.[4]

In Sunderland the pattern of industrial development provides an explanation for the processes of decay which have been examined. Confined within the narrow gorge of the River Wear, industry and port facilities have debouched into the coast area to the south of the river's mouth. The development of railways to serve this coastal industrial belt helps to elucidate the pattern of decay in Hendon. The Londonderry Railway, running from Seaham to Sunderland, was opened in 1854 and a station (which was closed in 1879) was opened at Hendon itself. The traffic generated by this railway, and the development of industry which it attracted to the Hendon area, changed the social value placed upon the area. Loss of rating value and the invasion of the area by lower income families—both of which would proceed westwards away from the expanding industrial area on the coast—would seem to be explained by this industrial development.

[1] R. A. Waters, *Hendon : past and present* (Sunderland, 1900), p. 6.
[2] *Ibid.* pp. 8 and 12.
[3] Hawley, *Human ecology*, p. 280.
[4] Davie, 'The pattern of urban growth', in G. P. Murdock (ed.).

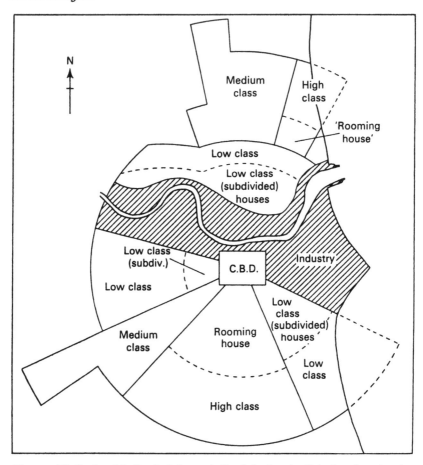

Fig. 3.23. Idealized model of ecological areas in Sunderland, 1963. Only the private housing areas are incorporated. Additional council housing areas are shown in Fig. 3.10.

MODELS OF URBAN GROWTH

To draw the strands of the argument together at this point, the development of the town's ecological structure which has so far been traced can be summarized. By superimposing the rating values and the degree of sub-divided houses in the private housing sector, a generalized scheme of the present-day ecology of the town can be visualized (Fig. 3.23). The Central Business District lies to the south of the river with industrial areas following both banks of the river and extending south along the coast to the south of the river mouth. The residential pattern in the north has some elements of the sectoral developments suggested by Hoyt (in that the highest rated area forms a partial sector adjacent to the amenity area of the seaboard), but in the main the northern area is in the form of a series of concentric rings of the Burgess type, progressing outwards from a poor, subdivided zone adjacent to the industrial area flanking the river, through a medium-rated area extending to the east–west railway line, and to a higher rated zone north of this and running to the boundary of the town. To the south of the river, the residential areas have developed a pattern closely akin to the sectoral model of Hoyt with four principal sectors: a low-class sector in the east which is highly subdivided at its northern apex; a highly rated sector next to it which reaches out from the inner areas in an expanding area to the southern boundary; a middle-class sector running out to the west; and finally, a second low-rated sector flanking the industrial areas of the river. As Hoyt's model would suggest, a rooming-house area has developed at the townward apexes of both of the highly rated sectors, one to the north of the river, the other, more extensive, to the south.

It is interesting to note the juxtaposition of sectoral development in one part of the town with concentric ring development in another. Other ecological studies have noted a similar tendency in towns both in the United States and Britain, without offering holistic explanation for this characteristic. Shaw, in one of the early writings of the classical school of Chicago, reluctantly acknowledged the existence of such pattern within Chicago itself.[1] With his zeal for the Burgess model, Shaw rejected the significance of those patterns of delinquency rates in the south of Chicago which did not accord with the Burgess concentric zones. Jones, too, finds a juxtaposition of sectors and zones in his work on Belfast. The area to the east of the River Lagan has developed concentric zones, while that to the west has developed sectors. Jones comments that for the moment, sectors and zones have to be laid aside as *complete* answers.[2] Again, if the data given for New Haven by Davie are plotted, a pattern similar to both Sunderland and Belfast emerges. In terms of occupation, income, the delinquency and dependency rates, and the social measures examined by Davie, a concentric ring of low status and high delinquency and dependency rates appears to the east of the river, and a pattern of distinct sectors radiating out from the Central Business District appears to the west.[3]

There are in fact many similarities in the forms which the residential patterns in all three of these examples have assumed, and they can help to offer a holistic explanation of the juxtaposition of sectors and concentric rings which have been noted. In each case, the town is river-based and the

---

[1] C. R. Shaw, *Delinquency areas* (Chicago, 1929).
[2] Jones, *A social geography of Belfast*, p. 274.
[3] Davie, *op. cit.* Smailes suggests that there are also somewhat comparable patterns within Greater London. Cf. A. E. Smailes, 'Greater London: the structure of a metropolis', *Geogr. Z.* LII (1964), 163–89.

industrial zones are found in close proximity to the river. The Central Business District lies close to, yet not adjacent to the river. Concentric zones are found in those residential areas which lie on the bank opposite to the Central Business District. Sectors are found in those residential districts which lie on the same side of the river as the Central Business District. What mechanisms can be suggested to account for these common features?

Two principal factors might be taken into account: attraction to the Central Business District and repulsion from industrial areas. The resolution of the balance of these forces in a river-based town gives rise to the juxtaposition of sectors and zones arranged in the manner that has been noted. The focus around which the residential areas of a medium-sized town are orientated is the Central Business District. This is the retail and entertainment centre which exerts much of the centripetal force which is responsible for the creation and maintenance of towns.[1] In the case of the professional workers, the Central Business District is also usually the place of work. We can thus see that the object of selecting a residential location within a town is to achieve maximum accessibility to this point while at the same time meeting other requirements such as good site, open land, access to the sea, high social value. Thus accessibility is the first of the two considerations. The second prime factor—avoidance of industrial areas—is one which, as has been noted, was stressed by Davie, and one which Hoyt recognized without stressing. Indeed the importance of the repulsive effects of industry on the pattern of invasion and the movement of the highly rated sector in the Hendon area, where it was seen to be responsible for the westward spread of decay and substitution of a concentric pattern by a sectoral one, has already been underlined. Especially in industrial towns like Sunderland (or Belfast, or New Haven, or Chicago) where industry is largely heavy and noxious, the patterning of the residential areas has to be viewed in terms of the location of industrial areas.

The interplay of these two forces can therefore be held to be primary in determining residential patterns. The highly rated area will tend to be located in such a position that it is at once accessible to the Central Business District and yet not adjacent to industrial areas. And indeed this offers an explanation for the juxtaposition of sectors and concentric zones on opposite sides of the river. On that side of the river on which the Central Business District has developed, access to it can be achieved directly without contact with the industrial areas of the river banks, whereas on the opposite side of the river direct access between the highly rated area and the Central Business District would involve building highly rated houses adjacent to the riverside industrial areas. To avoid this, a low-rated concentric zone is, as it were, allowed

[1] For an elaboration of these forces see C. C. Colby, 'Centrifugal and centripetal forces in urban geography', *Ann. Ass. Am. Geogr.* XXIII (1933), 1–20.

to interpose itself between the highly rated zone and the industrial area so that a concentric pattern develops which permits the maximum possible access between the highly rated area and the Central Business District without there being physical contact between highly rated housing and industrial areas. Thus for the highly rated areas on both sides of the river the two objectives have been achieved: in the one case by interposing the Central Business District itself between the highly rated area and industry; in the other, by using an intervening low-rated area as a buffer. The resolution of the balance of these two forces of attraction to the centre and repulsion from industry thus produces the juxtaposition of sectors and concentric zones on opposite sides of the industrial belt, which, in the case of Sunderland and the other towns cited, forms along the line of the river. The highly rated area forming a concentric ring on the opposite side of the river to the Central Business District might be expected to develop at a later date than the sector-wedge highly rated area, as indeed was the case in Sunderland.[1]

CONCLUSION

The Sunderland case confirms Jones' conclusion that neither the Burgess nor the Hoyt model can provide a total explanation of the patterns of residential segregation. The fault with such schemes lies in the emphasis which they place on the isolation of models *per se*, rather than the isolation of those forces which operate within the urban area to produce residential patterns whose precise form is conditioned by the interrelationship of a whole host of forces and factors. In this chapter an attempt was made merely to isolate two of what seem the most important factors in explaining the development of a particular town in the recent past: namely the centripetal/centrifugal forces and the repulsion exerted by industry on areas of highly rated housing. The analysis has however led some way to an understanding of the structure of Sunderland. It has also illustrated the usefulness of the little available documentary evidence in reconstructing a past geography of the town's social areas. The weaknesses of the analysis so far are twofold. First, only single variables have been considered. Rateable value or the subdivision of houses or a combination of both of these have been regarded as catch-all measures of the whole range of social and economic phenomena. The urban scene is much more complex than such a one-variable treatment might suggest. For example, given two areas of identical rateable value, the fact

---

[1] A most interesting urban model which is very similar to the pattern of sectors *and* zones which is advanced here has recently been suggested by Johnston. He combines the sectoral ideas of Hoyt with Sjoberg's concept that high status areas are found close to the centre of pre-industrial cities, to build up a combination of sectors and zones. Cf. R. J. Johnston, 'The location of high status residential areas', *Geogr. Annlr*, XLVIB (1966), 23–5.

that one has a predominantly young age structure while the other is over-whelmingly of older age structure is of vital importance in considering a host of social attributes in the two areas. Secondly, while it has been shown that the classical ecological models are diagnostic tools in the plotting of nineteenth-century urban patterns, and even for areas of twentieth-century private housing, they are incapable of assimilating the vast developments of local-authority housing which are so characteristic of most British towns. Most of the validity of these models in the present-day context is due to the fact that the present outlines of the private sector of the town are a palimpsest of the nineteenth-century developments. The council housing areas which have, of necessity, been ignored cover a large proportion of the total town. In 1961, 40 per cent of all dwellings were built or controlled by the local authority, and similar figures hold for other towns. Indeed, in 1961, as high a propor-tion as 40 per cent of all British towns of over 50,000 population had over 30 per cent of their population living in council houses. The appearance of such large areas of local authority housing has made nonsense of the rings or sectors of the classical ecological theory. Indeed, the game of hunt-the-Chicago-model seems to be exhausted so far as the analysis of modern developments in British urban areas is concerned. The development of local authority housing both in peripheral areas and in central redevelopment areas, together with the general increase in the role of central and local town planning, the spread of affluence and its concomitant eroding of social differentials, or at least the increase in social and geographical mobility that this has brought with it, and the very fact of the great decrease in the rates of urban growth: all have led increasingly to shortcomings in the classical models, and to their limited usefulness in the modern setting.[1]

In the following chapter the effects of the appearance of the large areas of council housing will be examined in closer detail and the reasons for neglect-ing them in the simple ecological analysis of this chapter will be elaborated. A more rigorous analytic technique will then be presented which can include a large number of variables and which can accommodate the council housing areas of the town, thus meeting the two most fundamental criticisms of a simple ecological analysis.

[1] Hoyt himself has recently discussed this point, recognizing the re-orientation which modern changes demand. Cf. H. Hoyt, 'Recent distortions of the classical models of urban structure', *Land Economics*, XL (1964), 199–212.

CHAPTER 4

MULTIVARIATE ANALYSIS

In the last chapter, classical models were used to analyse the growth and ecological development of residential areas in Sunderland and a single variable was used as an index of socio-economic conditions. Apart from the intrinsic value of this analysis, the approach was partly dictated by the lack of alternative information for dates before 1961, but there are two chief criticisms of this one-variable technique. First, the universe of data in urban areas is multi-faceted and its complexity cannot be subsumed under a single variable. Rating valuations are an insensitive indicator of any but social class factors and even here the association is by no means perfect. Second, recent changes in the structure of urban areas have largely invalidated many of the concepts of the classical ecologists and this is particularly the case with the appearance of local authority housing. In this chapter, therefore, the scope of the analysis will be widened not only by investigating the geographical association of more than one variable, but also by examining the changes which council housing has brought about and by attempting to apply a technique which can accommodate these changes and others which have characterized the modern period. First the simple associations between variables are studied to examine the effects of the introduction of council housing, and then the more complex patterns of association are examined by using the multivariate technique of component analysis.

The conclusions of the last chapter related entirely to the private housing areas of the town. It was possible in such areas to use rating valuations as a blanket index of social differentials. However, the existence of the large areas of council housing, which form a girdle of low status population around the outskirts of the town, was used to argue that such areas made nonsense of the classical models of urban growth. Looking now at a much wider selection of variables, it is possible to see in more detail the way in which council housing has also profoundly affected the whole basis of the internal structuring of the social composition of towns. For example, the inclusion of the council sector robs both rating valuations and the Jurors' index—two of the most commonly used indexes of social class—of much of their diagnostic value. If, first, the effect on rating values is examined, the census enumeration districts can be divided into three groups with respect to council housing:[1] those which include 90 per cent and more private houses (private); those which include 90 per cent and more council housing (council); and those which include less than 90 per cent of either type (mixed). The average rating values within each enumeration district can be correlated with the area's social class score as derived from the census socio-economic groups. The expected high positive association between rating values and social class in the private housing category has already been noted in justifying the use of rating values in the previous chapter.[2] However, this close association breaks down in the council sector as the following figures show:

Correlation between rating values and social class

(a) Private sector $r = 0.870$

(b) Mixed sector $r = 0.767$ } (d) Whole town $r = 0.606$

(c) Council sector $r = 0.210$

The coefficients for the private and mixed sectors are significant at the 1 per cent confidence level whereas that for the council sector is not significant even at the 5 per cent level. Thus, while social class may vary quite markedly from one council estate to another, rating values show little if any change and what variation does occur is not necessarily associated with the social class composition of the estate. By incorporating council areas into the overall analysis, it is obvious that the diagnostic value of rating values would have been seriously impaired since they have little, if any, meaning as indexes other than in the private housing sector.

The same is true of the use of the J-index, which has also been widely used as an index of social class. As in the case of rating values, there is a high positive association between the proportion of jurors and social class within the private housing sector ($r = 0.728$), but within the council sector there is little relationship, so that for the whole town the association falls to $r = 0.390$. While the distribution of jurors in the private sector corresponds closely with the maps of valuation patterns, in the council areas there is an inexplicable pattern of high and low proportions of qualified persons which bears no relation to the social class characteristics of the estates.

It can therefore be seen that the use of simple indexes of social class is no longer valid once the large areas of council housing are included in any analysis. While it appears that, once built, a certain amount of residential segregation and re-sorting does occur within local authority housing schemes,

[1] For rating values see references above, p. 103; for the use of the Jurors' index (J-index) see P. G. Gray *et al.*, *The proportion of jurors as an index of the economic status of a district* (Govt. Social Survey, London, 1951). Persons qualified for jury service are recorded in the annual *Register of Electors* for local areas.

[2] See above, p. 105.

in so far as land values no longer control such processes and certainly exert little control over the siting of such housing schemes, and also in that access to the Central Business District is no longer so operative a factor with the improvement of transport and the overriding dictates of the local authority itself, council areas have obviously had profound effects on the internal structure of urban areas.

In order to examine the more detailed effects of council housing on the whole range of socio-economic characteristics, it will be profitable to look at the patterns of correlation coefficients between a number of variables and to compare these patterns within the private housing areas of the town with those for the whole town when the areas of council housing have been in-cluded. Not only will this spotlight the disruptions caused by council housing, but it will also provide a useful means of studying the simple associations which exist between all those variables which were selected to form the material for the later component analysis.

In all, a total of 30 variables was used. Of these, 27 were taken from census material and their selection was guided by the attempt to provide a comprehensive coverage of a variety of social and economic traits. As Table 4.1 shows, variables cover the social composition of the population, its age structure, household tenure, household accommodation and composition, and the nature of the housing type. The individual variables each have a certain degree of theoretical support for their selection in that they are parameters of aspects of social structure and areal differentiation which previous research has shown to be of some importance. In addition to these census measures, three variables were drawn from the Valuation Records and the Register of Electors: median gross rateable value, the percentage of subdivided houses and the percentage of persons qualified for jury service. These were included since they are widely used indexes in social research and their comparison with the fuller material of the census was thought to be an interesting test of their validity, as well as of the use of rateable values in the previous chapter. The areal units which are used in studying these 30 variables are the census enumeration districts into which the town was divided for the 1961 census. A total of 263 enumeration districts was used and the non-census variables were calculated for the areal framework of these districts. In many ways, the enumeration districts are not ideal areal units since, being delimited with an eye to administrative convenience rather than internal homogeneity (unlike the rating areas of the last chapter), some of them include differing types of area. There is, for example, no consistent differentiation of private and council areas. Nevertheless, since the segrega-tion of residential areas within the town is so well developed and since the enumeration districts are so small, this problem of intra-area heterogeneity affects only a very small number of areas and attention is drawn to it in the few cases where it affects the subsequent analysis.

Product-moment correlation coefficients were calculated between each

TABLE 4.1. *Variables used in component analysis*

I *Social composition*

1 Percentage class I—occupied and retired males in socio-economic groups 1, 2, 3, 4 and 13 (professional, employers and managers).
2 Percentage class II—males in s.e.g.s 5 and 6 (non-manual workers).
3 Percentage class III—males in s.e.g.s 8, 9, 12 and 14 (foremen, skilled manual, workers on own account).
4 Percentage class IV—males in s.e.g.s 7, 10, 11, 15, 16 and 17 (personal service, semi-skilled manual, unskilled manual, agricultural, armed services, inadequately described).
5 'Class score' (derived by simple weighting of above groups: group 1 weighted by 4, group 2 by 3, etc.).
6 Percentage of population aged over 16 with terminal education age below 16.
7 Females in the labour force as percentage of females aged 15 and over.
8 Percentage of women aged 20–24 ever married.
9 Percentage of persons aged over 21 qualified for jurors' service (*J*-index).

II *Age structure*

10 Percentage of persons aged 0–14.
11 Percentage of persons aged 20–24.
12 Percentage of persons aged 65 and over.
13 Fertility ratio 1—persons aged 0–4 as percentage of females aged 15–44.
14 Fertility ratio 2—persons aged 0–9 as percentage of females aged 15–44.

III *Household tenure*

15 Percentage of households owner occupied.
16 Percentage of households renting accommodation from local authority.
17 Percentage of households renting private furnished accommodation.
18 Percentage of households renting private unfurnished accommodation.

IV *Household accommodation and composition*

19 Percentage of 1-person households.
20 Percentage of 2-person households.
21 Percentage of 6-person households or larger.
22 Average number of persons per room.
23 Percentage of households at densities of over 1·5 persons per room.

V *Housing characteristics*

24 Median gross rateable value.
25 Percentage of houses with over 8 rooms.
26 Percentage of unsubdivided houses.
27 Percentage of households with one household space.
28 Percentage of households sharing dwellings.
29 Percentage of households without a fixed bath.
30 Percentage of households with exclusive use of all four amenities (i.e. cold running water, hot running water, water closet, fixed bath).

SOURCES. *i* Variable 9—*Register of Electors*, Sunderland C.B., 1963.
ii Variables 24 and 26—*Valuation Lists* (Sunderland, 1963).
iii Variables 1 to 7—1961 census, 10 per cent sample, enumeration district data, supplied privately.
iv Remaining variables—1961 census, 100 per cent census, enumeration district data, supplied privately.

pair of these 30 variables, and to examine the effects of council housing on the internal structure of the town the dichotomy between private and council areas will be pursued further by looking initially at the matrix of coefficients for the private areas alone. For this purpose the distinction between 'private' and 'council' is made between those enumeration districts which include less than 25 per cent of council houses and which are con- sidered to be 'private' areas, and those with 25 per cent and more council houses which are considered to be 'council'. In practice there is not a great deal of admixture of private and council housing since the two are generally highly segregated in space. It is largely in the inner areas of the town where redevelopment has taken place that much admixture of the two occurs. Of the total of 263 enumeration districts, only 43, or 16 per cent, had more than 10 per cent but less than 90 per cent of either private or council houses within them. The remaining 84 per cent of enumeration districts had over 90 per cent of the dominant type, either private or council. By and large, therefore, the two types of housing were quite distinct. Using this definition, the enumeration districts were divided into 159 'private' areas and 104 'council' areas.

CORRELATION MATRIX: PRIVATE HOUSING

The values of the Pearsonian product-moment correlation coefficients of the 30 variables for the 159 private areas are shown in the upper segment of the matrix in Appendix B.[1] With 157 degrees of freedom, coefficients of ± 0.201 and over are significant at $P = 0.01$. To simplify the analysis of the matrix, however, coefficients of ± 0.500 and over have been regarded as 'high' coefficients and their distribution is shown in Table 4.2. The distribution of such 'high' scores illustrates how much more diagnostic certain of the variables are than others within the total universe of data. Each variable has an individual possible maximum of 29 'high' scores whereas the highest number actually found for any variable is 18 (which is found in the two variables persons per room and the percentage of households renting private furnished accommodation). To this extent, these two variables appear to be discriminating measures of a range of socio–economic data within the private housing sector of the town. Other variables with large numbers of 'high' scores are as follows: the percentage of owner occupiers, the percentage of households with all four amenities (each of which has 17 'high' scores), the median gross rateable value, the percentage of households without exclusive use of fixed bath (each of which has 16 'high' scores), the average class scores, and the percentage of unsubdivided houses (each of which has 15 'high'

---

[1] For some cautionary points relating to such areal correlations, see below, pp. 155–9.

TABLE 4.2. 'High' coefficients in the matrix for private housing (159 e.d.s.)

Variables		1	2	3	4	5	6	7	8	9	10	11	12	13	14	15	16	17	18	19	20	21	22	23	24	25	26	27	28	29	30
Social class	1	*	·	-	-	+	-	·	·	+	·	·	·	·	-	+	+	-	-	·	·	·	-	-	+	·	·	·	·	-	+
	2	·	*	-	-	+	-	·	·	+	·	·	·	·	-	+	+	·	·	·	·	·	-	-	+	·	·	·	·	-	+
	3	-	-	*	·	-	+	·	·	·	·	·	·	·	·	·	·	·	·	·	·	·	·	+	-	·	·	·	·	·	·
	4	-	-	·	*	-	+	·	·	-	·	·	·	·	+	-	·	+	+	·	·	+	+	+	+	·	+	·	·	·	-
	5	+	+	-	-	*	-	·	·	+	·	·	·	·	-	+	·	·	·	·	-	-	+	+	-	·	-	·	·	+	-
	6	-	-	+	+	-	*	·	·	-	·	·	·	·	·	·	·	·	·	·	+	+	·	·	·	·	·	·	·	·	·
	7	·	·	·	·	·	·	*	·	·	·	·	·	·	·	·	·	·	·	·	·	·	·	·	·	·	·	·	·	·	·
	8	·	·	·	·	·	·	·	*	·	·	·	·	·	+	·	·	+	+	·	-	-	-	-	·	·	·	·	·	-	-
	9	+	+	·	-	+	-	·	·	*	+	·	·	·	+	+	+	·	+	·	-	+	-	+	+	+	-	+	+	-	+
Age structure	10	·	·	·	·	·	·	·	+	+	*	-	·	·	·	-	·	·	-	·	·	-	-	-	-	+	·	·	·	·	-
	11	·	·	·	·	·	·	·	·	·	·	*	+	·	+	·	·	·	·	·	-	+	+	·	·	·	·	·	·	·	·
	12	·	·	·	·	·	·	·	·	·	-	+	*	·	·	·	·	·	·	·	·	·	·	·	·	·	·	·	·	·	·
	13	·	·	·	·	·	·	·	·	·	·	·	·	*	·	·	·	·	·	·	·	·	·	·	·	·	·	·	+	·	·
	14	-	-	·	+	-	·	·	+	+	·	+	·	·	*	-	·	+	+	·	+	+	+	+	-	+	·	+	+	+	-
Tenure	15	+	+	·	-	+	·	·	·	+	-	·	·	·	-	*	·	·	-	·	+	-	-	·	+	+	+	·	·	-	+
	16	+	+	·	·	·	·	·	·	+	·	·	·	·	·	·	*	·	·	·	·	·	·	·	·	·	·	·	·	·	·
	17	-	·	·	+	·	·	·	+	·	·	·	·	·	·	·	·	*	+	·	-	+	+	·	·	·	·	·	+	·	-
	18	-	·	·	+	·	·	·	+	+	-	·	·	·	+	-	·	+	*	·	-	+	+	+	-	-	·	·	+	+	+
Household composition	19	·	·	·	·	·	·	·	·	·	·	·	·	·	·	·	·	·	·	*	·	·	·	·	·	·	·	·	·	·	·
	20	·	·	·	·	·	+	·	-	-	-	+	·	·	-	+	-	·	-	·	*	-	-	·	-	·	·	·	·	·	-
	21	·	·	·	+	·	+	·	-	+	·	+	·	·	+	-	·	+	+	·	-	*	+	+	·	·	·	·	·	+	+
	22	-	-	·	+	·	·	·	-	-	-	+	·	·	+	-	·	+	+	·	-	+	*	+	-	·	·	·	·	+	+
	23	-	-	+	+	·	·	·	-	+	·	·	·	·	+	·	·	·	+	·	·	+	+	*	-	·	·	·	·	+	+
Housing characteristics	24	+	+	-	+	-	-	·	·	+	-	·	·	·	-	+	·	·	-	·	-	·	-	-	*	+	·	+	-	-	+
	25	·	·	·	·	·	·	·	·	+	+	·	·	·	+	+	·	·	-	·	·	·	·	·	*	-	+	·	+	+	+
	26	·	·	·	+	·	·	·	·	-	·	·	·	·	·	+	·	·	·	·	·	·	·	·	·	*	+	·	+	·	·
	27	·	·	·	·	·	·	·	·	+	·	·	·	·	+	·	·	·	·	·	·	·	·	·	+	*	+	·	·	·	·
	28	·	·	·	+	·	·	·	·	+	·	·	·	+	+	·	·	+	+	·	·	·	·	·	-	+	·	*	-	·	-
	29	-	-	·	+	+	·	·	-	-	·	·	·	·	+	-	·	·	+	·	·	+	+	+	-	+	·	*	-	·	-
	30	+	+	·	-	-	-	·	-	+	-	·	·	·	-	+	·	-	+	·	-	+	+	+	+	+	·	·	-	-	*

Number of coefficients ≥ ±0·500: | | 12 | 13 | 5 | 14 | 15 | 13 | 0 | 6 | 14 | 11 | 0 | 1 | 0 | 14 | 17 | 0 | 1 | 18 | 3 | 7 | 6 | 18 | 17 | 16 | 1 | 15 | 2 | 4 | 16 | 17 |

NOTES. Correlation coefficients: + 0·500 and greater   − −0·500 and greater   * self-correlation scores). By comparison, a number of variables have no 'high' scores at all. This is the case with the proportion of females in the labour force, the percentage of people aged 20–24, one of the fertility ratios and, understandably, the percentage of households living in council property. These variables therefore appear to have little diagnostic value in the private housing sector.

The overall pattern of 'high' correlations shows a wide scatter throughout most of the matrix. To help interpret the matrix, Table 4.2 has been divided into five rough groupings of related measures which, while by no means rigid, provide approximate clusters of similar variables. Apart from the expected 'high' correlations within these groupings, which is especially marked within the social class, household composition and housing characteristic groups, there are strong associations between a number of variables in different groups. This is particularly true in the case of variables from the social class and the housing characteristic groups. The class score, for example, correlates highly with the percentage of households lacking baths ($r = -0.754$), and the percentage of people in social class IV correlates highly with the percentage of people living at over $1\frac{1}{2}$ persons per room ($r = 0.705$). There are also strong associations between the age characteristics and the household composition and household tenure groups. Persons aged 0–14, for example, is closely associated with 6-person households ($r = 0.518$), with 2-person households ($r = -0.687$), and with the average number of persons per room ($r = 0.729$). The tenure and household composition groups are also strongly related. For example, the percentage of owner occupiers is strongly related positively with 2-person households and negatively with 6-person households, and the reverse associations are found between the percentage of private households rented unfurnished and these two composition variables. Finally, the tenure group is strongly related to the housing characteristic group. The percentage of privately rented unfurnished accommodation has high coefficients with the percentage of households with all four amenities ($r = -0.785$), with the percentage of unsubdivided dwellings ($r = -0.797$), and with the average number of persons per room ($r = 0.763$).

The pattern which emerges therefore shows a large number of very high coefficients with a great deal of interconnection between variables drawn from most of the rough groupings which have been delineated. It is also interesting to note the pattern of correlations of the two 'classical' indexes of social class—rateable values and the *J*-index. Rateable value has a large number of high coefficients. It is strongly correlated positively with social class, with owner-occupancy, and with good housing amenities. It is strongly associated negatively with overcrowding. Its associations with the age structure are only slightly less pronounced since it has a moderately high negative correlation with a young age structure ($r = -0.400$ with the percentage of persons aged 0–14 and -0.399 and -0.556 with the two fertility ratios). Furthermore, relating to the use of subdivided housing in the last chapter, its association with subdivided housing is strongly negative ($r = -0.425$ with the proportion of shared households and 0.681 with the percentage of un-subdivided housing). The *J*-index, which itself is closely correlated with rateable values, shows a very similar, although less well-established, pattern of correlations. The overall pattern of the individual correlations of these two variables therefore appears to vindicate the use which was made of gross rateable value in the last chapter in analysing the private housing sector.

CORRELATION MATRIX: WHOLE TOWN

With the inclusion of council housing, however, the pattern changes dramatically. Throughout the last chapter and in much of the present chapter, great stress has been laid on the disrupting effects of local authority housing on the traditional methods of analysing urban social structure. The comparison of the private housing matrix and the matrix for the whole town, which includes council housing, makes it possible to underline and highlight these changes in a more precise and detailed fashion.

By including the council areas, the total number of units (enumeration districts) in the analysis rises from 159 to 263 and the level of significance for the matrix of coefficients for the whole town (which will be referred to briefly as the 'total matrix') consequently changes. Whereas in the private matrix the 1 per cent confidence level was ± 0.201 and over, in the total matrix, with 261 degrees of freedom, it becomes ± 0.158 and over. With this lower confidence level, the value of 'high' coefficients in the total matrix, which corresponds to ± 0.500 in the private matrix, is ± 0.409 and over.[1] By using

---

[1] This value, equivalent to ± 0.500 in the private matrix, is derived from the formula used for calculating Student's *t*. Since, with the large number of degrees of freedom, the distributions involved here are virtually normal, *t* values are practically identical for the two sets of data. By re-expressing the formula for *t* in terms of *r*, we get the following

$$t = \sqrt{\frac{r^2(N-2)}{1.0 - r^2}} \quad (1) \qquad \text{becomes} \qquad r = \sqrt{\frac{t^2}{t^2 - (N-2)}} \quad (2).$$

Thus, to find the equivalent in the total matrix of ± 0.500 in the private matrix, we calculate (formula (1)):

$$t^2 = \frac{0.5 \times 157}{1.0 - 0.5} = 39.25,$$

and substitute this value of *t* in formula (2) for the total matrix as follows:

$$r = \sqrt{\frac{39.25}{39.25 - 261}} = 0.409.$$

Similarly calculated equivalent values of *r* in the private and total matrices are shown diagrammatically in Fig. 4.2.

TABLE 4.3. 'High' coefficients in the matrix for the whole town (263 e.d.s.)

Variables	1	2	3	4	5	6	7	8	9	10	11	12	13	14	15	16	17	18	19	20	21	22	23	24	25	26	27	28	29	30
Social class																														
1	*	+	+	−	+	+	·	·	·	+	·	·	·	·	+	+	·	·	·	·	·	·	·	+	·	·	·	·	·	·
2	+	*	−	−	+	−	·	·	·	·	·	·	·	−	+	+	·	·	·	·	−	−	−	+	+	·	·	·	·	·
3	+	−	*	−	·	+	·	·	·	·	·	·	·	·	·	·	·	·	·	·	·	·	·	·	·	·	·	·	·	·
4	−	−	−	*	+	+	·	·	·	·	·	·	·	·	−	−	·	·	·	·	·	·	·	·	·	·	·	·	·	·
5	+	+	·	+	*	·	·	·	·	·	·	·	·	·	+	+	·	·	·	·	·	−	·	+	·	·	·	·	·	·
6	+	−	+	+	·	*	*	·	·	·	·	·	·	·	·	·	·	·	·	·	·	·	·	·	·	·	·	·	·	·
7	·	·	·	·	·	*	*	·	·	·	·	·	·	·	·	·	·	·	·	·	·	·	·	·	·	·	·	·	·	·
8	·	·	·	·	·	·	·	*	*	·	·	·	·	·	·	·	·	−	·	·	·	·	·	−	·	·	·	·	+	+
9	·	·	·	·	·	·	·	*	*	·	·	·	·	·	·	·	·	+	+	·	−	·	+	+	·	·	·	·	·	+
Age structure																														
10	+	·	·	·	·	·	·	·	·	*	*	·	·	·	+	+	·	·	−	−	−	·	·	*	−	·	−	·	·	−
11	·	·	·	·	·	·	·	·	·	*	*	·	·	·	·	·	·	·	+	·	·	·	·	·	·	·	·	·	·	·
12	·	·	·	·	·	·	·	·	·	·	·	*	*	·	·	−	·	+	+	·	·	·	·	·	·	·	·	·	+	·
13	·	·	·	·	·	·	·	·	·	·	·	*	*	·	·	·	·	·	·	·	·	·	·	·	·	·	·	·	·	·
14	·	−	·	·	·	·	·	·	·	·	·	·	·	*	·	·	·	·	−	·	·	·	+	·	·	·	·	·	·	·
Tenure																														
15	+	+	·	−	+	·	·	·	·	+	·	+	·	·	*	+	·	+	+	+	−	+	·	−	−	·	−	+	+	+
16	+	+	·	−	+	·	·	·	·	+	·	+	·	·	+	*	*	·	−	*	*	*	·	·	−	·	·	−	·	−
17	·	·	·	·	·	·	·	·	·	·	·	·	·	·	·	*	*	+	·	+	·	·	·	·	−	−	·	−	·	−
18	·	·	·	−	·	·	·	−	+	·	+	+	·	·	+	·	*	*	−	·	·	·	·	·	·	·	·	−	+	−
Household composition																														
19	·	·	·	·	·	·	·	·	·	−	+	+	·	−	+	−	·	−	*	+	−	+	·	*	+	·	+	+	+	+
20	·	·	·	·	·	·	·	·	+	−	·	+	·	·	+	*	+	·	+	*	+	+	·	·	·	·	−	−	·	·
21	·	−	·	·	·	·	·	·	+	−	·	−	·	·	−	*	·	·	−	+	*	+	+	·	·	−	·	·	−	·
22	·	−	·	·	−	·	·	·	·	·	·	−	·	+	+	*	·	·	+	+	+	*	+	·	−	−	·	+	·	·
23	·	−	·	·	·	·	·	·	+	·	·	−	·	+	·	·	·	·	·	·	+	*	*	·	·	·	+	+	·	·
Housing characteristics																														
24	+	+	·	·	·	·	·	−	+	·	·	·	·	·	−	·	·	−	*	·	·	·	·	*	·	+	+	+	+	+
25	·	+	·	·	·	·	·	·	·	−	·	·	·	·	+	−	−	·	+	·	·	−	·	·	*	+	·	*	+	+
26	·	·	·	·	·	·	·	·	·	·	·	·	·	·	·	·	−	·	·	·	−	−	·	+	+	*	−	+	+	+
27	·	·	·	·	·	·	·	·	·	·	·	·	·	·	−	·	·	·	+	−	·	·	+	+	·	−	*	−	−	−
28	·	·	·	·	·	·	·	·	·	·	·	·	·	·	+	−	−	−	+	−	·	+	+	+	*	+	−	*	+	*
29	·	·	·	·	·	·	·	+	·	·	·	+	·	·	+	·	·	+	+	·	−	·	·	+	+	+	−	+	*	+
30	·	·	·	·	·	·	·	+	+	−	·	·	·	·	+	−	−	−	+	·	·	·	·	+	+	+	−	*	−	*
Number of coefficients ≧ ± 0·409:	9	13	4	9	14	12	0	9	8	11	4	9	5	9	14	13	7	13	13	11	11	14	14	14	3	13	6	10	12	14

NOTES. Correlation coefficients: + 0·409 and greater   − −0·409 and greater   * self-correlation this lower level the comparison of coefficients in the two matrices is more truly valid. The actual coefficients for the total matrix are shown in the lower segment of the matrix in Appendix B, while the simplified pattern of 'high' coefficients is shown in Table 4.3. With this lower level of exactly comparable 'high' coefficients the total number of 'high' coefficients in the total matrix shows a slight rise. Whereas there were 276 'high' values in the private matrix, there are 298 in the total matrix. This slight rise in the overall total of 'high' coefficients, however, conceals the fact that the inclusion of the council housing areas greatly reduces the value of the great majority of coefficients. Fig. 4.1 compares the frequency distributions of the coefficients

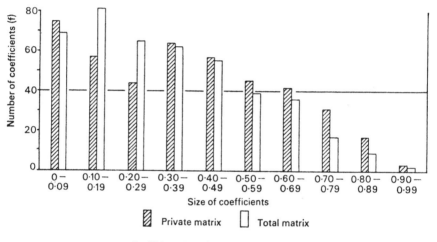

Size of coefficients

Private matrix   Total matrix

Coefficient signs (+ and −) have been ignored

Fig. 4.1

Frequency distributions of correlation coefficients in the private and the total matrices

in the private matrix and the matrix for the whole town. It is apparent that those in the private matrix have a much greater spread of values with a large number of very high coefficients. By comparison with the matrix for the whole town, the private matrix has consistently more coefficients in all of the higher value ranges above ±0·300, and slightly more which are uncorrelated in the range ±0·000 to ±0·099. The inclusion of the council housing thus has the effect of lowering the value of the associations. This can also be seen in the reduction of the average values of coefficients for each of the 30 variables (Table 4.4). While a few of these average coefficients show a rise in value, 21 out of the 30 show a fall in value so that the range of values in the private matrix is from 0·068 to 0·549 (ignoring positive and negative signs) as against the range of the matrix for the whole town which is from 0·135 to

TABLE 4.4. *Changes between the private and the total matrices*

		Number of 'high' coefficients			Average coefficients				
		(a) No. in private matrix	(b) No. in total matrix	(c) Difference (b) from (a)	(d) Private matrix	(e) Total matrix	(f) Rank of (d)	(g) Rank of (e)	(h) Difference (g) from (f)
	Variable								
1	Class I	12	9	− 3	0·420	0·346	14	18	− 4
2	Class II	13	13	0	0·411	0·343	15	19	− 4
3	Class III	5	4	− 1	0·263	0·228	24	29	− 5
4	Class IV	14	9	− 5	0·453	0·328	13	20	− 7
5	Class score	15	14	− 1	0·510	0·409	7	3	+ 4
6	Terminal education age below 16	13	12	− 1	0·463	0·383	10	9	+ 1
7	Females working	0	0	0	0·068	0·135	30	30	0
8	Females married	6	9	+ 3	0·345	0·281	18	23	− 5
9	J-index	14	8	− 6	0·457	0·350	11	17	− 6
10	Aged 0–14	11	11	0	0·401	0·357	16	16	0
11	Aged 20–24	0	4	+ 4	0·238	0·239	25	28	− 3
12	Aged 65 and over	1	9	+ 8	0·233	0·369	26	11	+15
13	Fertility ratio (0–4)	0	5	+ 5	0·313	0·271	22	26	− 4
14	Fertility ratio (0–9)	14	9	− 5	0·455	0·314	12	21	− 9
15	Owner occupiers	17	14	− 3	0·549	0·375	1	10	− 9
16	Council	0	13	+13	0·079	0·434	29	1	+28
17	Renting furnished	1	7	+ 6	0·176	0·281	27	24	+ 3
18	Renting unfurnished	18	13	− 5	0·507	0·391	8	7	+ 1
19	1-person households	3	13	+10	0·267	0·364	23	13	+10
20	2-person households	7	11	+ 4	0·351	0·387	17	8	+ 9
21	6-person households	6	11	+ 5	0·331	0·365	20	12	+ 8
22	Persons per room	18	14	− 4	0·525	0·394	4	5	− 1
23	Over 1½ per room	17	14	− 3	0·544	0·364	2	14	− 12
24	Rateable value	16	14	− 2	0·527	0·408	3	4	− 1
25	Houses over 8 rooms	1	3	+ 2	0·160	0·240	28	27	+ 1
26	Unsubdivided	15	13	− 2	0·483	0·359	9	15	− 6
27	One household space	2	6	+ 4	0·326	0·274	21	25	− 4
28	Households sharing	4	10	+ 6	0·344	0·297	19	22	− 3
29	Minus bath	16	12	− 4	0·514	0·393	6	6	0
30	All four amenities	17	14	− 3	0·523	0·410	5	2	+ 3

NOTES.
i 'High' coefficients $\begin{cases} \text{in private matrix} & \geqq \pm 0{\cdot}500 \\ \text{in total matrix} & \geqq \pm 0{\cdot}409 \end{cases}$

ii Average coefficients: for each variable $\dfrac{\Sigma r}{N}$ (ignoring the sign).

0·434. It would therefore seem that the inclusion of the council areas has had the effect of reducing the high coefficients and raising the very low. This adds confirmation to the earlier suggestions that council housing blurs many of the differentials in urban areas and makes for greater overall uniformity.

In examining the matrix for the whole town, two main aspects will be considered: first the general pattern of correlations and second the detailed changes which have been brought about through the introduction of the council sector.

First the distribution of 'high' scores might be examined. Table 4.4 shows that the variables have changed in their diagnostic value. The largest individual number of 'high' scores falls from eighteen to fourteen, which is found in six of the variables: the class score, owner-occupiers, persons per room, density of over $1\frac{1}{2}$ per room, rateable value and the presence of all four amenities. Of these, the average number of persons per room stands out as by far the most highly diagnostic having no fewer than twelve coefficients as high as ± 0.600 and over. Both in the private sector and in the matrix for the whole town, this variable emerges as being of great discriminatory value. The only variable which has no 'high' scores is the percentage of females in the labour force, which likewise had none in the private matrix.

The average correlations of each variable may be used as a shorthand estimate of the pattern of change between the two matrices. Table 4.4 lists these and shows that the social class variables have consistently fallen in rank with the inclusion of the council sector. So too have the housing characteristic variables. Large rises in rank, on the other hand, are found in the percentage of council tenants which rises, understandably, from second bottom to top in terms of its average coefficient. More interesting are the rises in the percentage of persons aged 65 and over, the percentage of 1-person households, 2-person households and 6-person households. These average coefficients however are no substitute for a detailed examination of the changes which have occurred in the actual values of individual coefficients since they conceal differences in the distribution of high and low scores within any one variable, as in the case, for example, of the variable measuring persons living at over $1\frac{1}{2}$ persons per room. The actual changes will therefore be examined in more detail to identify more precisely the effects of the inclusion of council housing.

Comparing the two patterns of 'high' scores in the matrices (Tables 4.2 and 4.3), it can be seen that the close relationship between social class and the physical nature of housing has been lost in the matrix for the whole town and that clusters of high scores are found instead, first, concentrated much more in the five intra-group correlations, and second, between the variables in the age structure, tenure, and household composition groups, between which there are strong correlations. The loss of correlation between class and physical amenities is a direct consequence of the improved housing facilities of the council houses. It is suggested, for example, by the fall in the coefficients between the social class score and the presence of all four household amenities where the coefficient falls from $r = 0.732$ in the private matrix to

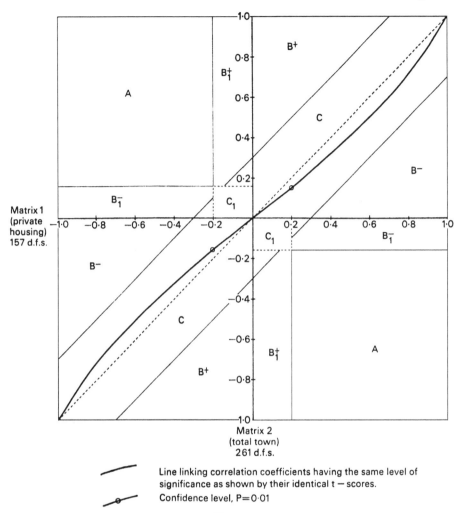

Fig. 4.2

Relationship between correlation coefficients in the private matrix and the total matrix

Coefficient values in the private matrix are shown on the *x* axis; those in the total matrix are shown on the *y* axis. The dashed diagonal joins identical values; the heavy line joins coefficients having the same level of significance.

The meaning of the various spaces on the graph is as follows:

Space *A*—coefficients which have changed the direction of their sign, with both coefficients being significant at $P = 0.01$.

Space *B*—coefficients which have risen (+) or fallen (–) in value by 0.3 or more from the private to the total matrix. (B_1 is a variant involving a non-significant change of sign.)

Space *C*—coefficients which have changed in value by less than ±0.3. (C_1 is a variant involving non-significant change of sign.)

$r = 0.199$ in the matrix for the whole town. Likewise it is seen in the actual reversal of the association between owner occupancy and amenities where, in the private matrix, the association is highly positive ($r = 0.799$) while in the matrix for the whole town it is non-significant ($r = -0.026$). In the private sector of the town, in other words, owner occupancy is closely related to better housing amenities, but when the relatively well-equipped council housing is included owner occupancy is no longer exclusively associated with good housing. To some extent, this change is also associated with the tendency for young married couples to buy houses which had previously been rented. This tendency has accelerated since 1945, reflecting growing affluence, and causing the marked contraction of the private rented housing sector which has been the cause of much recent concern.[1] The houses which are now included within the owner-occupied category therefore include many old and poorly equipped dwellings which would previously have been privately rented.

Patterns of change as between the two matrices

The complex changes such as those mentioned above can be clarified and summarized if we consider the graph of Fig. 4.2. In this graph the relationship between coefficients in the two matrices is illustrated graphically, with the line linking coefficients having the same level of significance plotted around the diagonal. If the coefficients in each matrix were plotted on this graph they would fall in one of a number of types of space. Taking the accepted level of significance at $P = 0.01$, the first is space A in which coefficients have changed in their direction from positive to negative, or vice versa, with both coefficients being significant. Second is space B in which the coefficients, while having the same sign, have either increased or decreased in value by over 0.300 when the council housing areas are included.[2] The space B_1 is a variant of this, in which the change of value has been equally large, but at the same time there has been a change of sign and a change from a non-significant to a significant value, or vice versa. Third is space C in which any change between the two matrices has been less than 0.300.

The two most interesting and pertinent of these changes are those included

[1] For the contraction of the privately rented housing sector at a local level see Cullingworth, *Housing in transition, op. cit.* At a national level see *The housing programme, 1965–70*, Cmnd. 2838 (London, 1965), which emphasizes the spread of owner-occupancy at the expense of privately rented housing.

[2] The value ± 0.300 is selected arbitrarily as representing a large amount of change in value. As Fig. 4.2 shows, however, it is statistically meaningful in that at no point does the line relating equally significant correlation coefficients cross this level of ± 0.300 from the diagonal. It thus represents an over-generous estimate of large significant changes in the values of coefficients as between the two matrices.

TABLE 4·5. Changes in coefficient values as between the private and total matrices

Variables	1 2 3 4 5 6 7 8 9 10 11 12 13 14 15 16 17 18 19 20 21 22 23 24 25 26 27 28 29 30
Social class	1–9
Age structure	10–14
Tenure	15–18
Housing composition	19–23
Household characteristics	24–30

	1	2	3	4	5	6	7	8	9	10	11	12	13	14	15	16	17	18	19	20	21	22	23	24	25	26	27	28	29	30
No. of increases:								3	3	3	3	2			2	2	1	1	6	2	2			2	1			1		1
No. of decreases:	5	5	3	4	6	5	1	9	1	2	6	6	1	4	13	11		1	3	3	8						4	4	10	2
No. of significant changes of sign:	5		3		5		1	1	7	3	1	1	1	1	1	1	3	3	6	6		2					3	1	2	3

Appropriate space in Fig. 4.2
A
B⁺ or B₁⁺
B⁻ or B₁⁻

NOTES.
□ Significant change of sign (sign changes: in private matrix $r \geqq \pm 0\cdot201$, in total matrix $r \geqq \pm 0\cdot158$)
+ Value of r increases by 0·3 or more in the total matrix as compared to the private matrix
− Value of r decreases by 0·3 or more in the total matrix as compared to the private matrix
* Self-correlation.

147

within spaces *A* and *B*, in which there has been, first, a significant change of sign or, second, a large change in the value of the two coefficients as between the two matrices. The changes which are effected by including council housing can therefore be summarized by examining those variables which are involved in such changes, and they are shown in Table 4.5.

TABLE 4.6. *Changes of sign between correlation matrices* (in which both coefficients are significant at $P = 0.01$)

					Correlation coefficients	
	Variable			Variable	Private matrix	Total matrix
10	Percentage aged 0–14 with	{	11	Percentage aged 20–40	0·325	−0·291
			18	Percentage rented unfurnished	0·504	−0·205
			30	Percentage with all four amenities	−0·426	0·218
11	Percentage aged 20–24 with		14	Fertility ratio (0–9)	0·378	−0·220
19	Percentage 1-person households with		15	Percentage owner occupiers	−0·453	0·229
		⎧	8	Percentage females ever married	−0·252	0·345
			9	*J*-index	0·291	−0·308
20	Percentage 2-person households with		11	Percentage aged 20–24	−0·250	0·180
			18	Percentage rented unfurnished	−0·569	0·289
			29	Percentage minus exclusive bath	−0·357	0·280
		⎩	30	Percentage with all four amenities	0·332	−0·354
		⎧	13	Fertility ratio (0–4)	0·308	−0·185
			18	Percentage rented unfurnished	0·589	−0·326
21	Percentage 6-person households with		26	Percentage undivided dwellings	−0·372	0·166
			28	Percentage households sharing	0·252	−0·190
			29	Percentage minus exclusive bath	0·241	−0·343
		⎩	30	Percentage with all four amenities	−0·238	0·395

First, those cases in which there has been a significant change of sign will be examined. In all, there are 17 sets of such significant changes and the values of the coefficients concerned are shown in Table 4.6. Much of the change can be seen to be associated with the household size variables: six of the changes relate to the percentage of 2-person households. This is a reflection of the unbalanced age structure of the council house intake which, concentrating on large families with many children, includes few old persons living alone or in 2-person households. This obviously has its effects upon the correlation coefficients. For example, whereas in the private housing matrix the association between privately rented unfurnished accommodation and both 1-person households and 2-person households was strongly negative ($r = -0.453$ and $r = -0.569$), with the inclusion of the council sector with its low proportion of old persons, both associations become positive ($r = 0.229$ and $r = 0.289$ respectively). Similarly, while 2-person households were

148

positively associated with good housing amenities in the private sector, in the matrix for the whole town they are negatively associated. In other words, within the private housing sector considered separately, 2-person households are concentrated in housing with relatively good facilities, but within the town as a whole, and thus including the better-appointed council housing, the 2-person households are concentrated in housing of relatively poor quality. Obviously the housing itself does not alter, nor does the location of 2-person households, but the total household population which is used as a yardstick does alter and changes the pattern of the associations. This serves to underline the nature of the council housing and its effects upon the overall structural pattern.

The same type of changes can be seen in the associations with 6-person households. Here the tendency for selectivity to operate within council housing policies again has the effect of altering many of the associations found in the private matrix. For example, whereas 6-person households in the private matrix are positively associated with privately rented unfurnished accommodation, with shared dwellings, with high fertility ratios and with poor housing facilities, in the matrix for the whole town they are seen to be negatively associated with these characteristics since the council areas in which they are concentrated (the association between the two is $r = 0.727$) are areas of un-subdivided and well-equipped housing.

Secondly, if the analysis turns from the actual changes of signs to the cases of large increases or decreases in the values of the coefficients, further revealing light is thrown on the nature and effects of council housing areas. Table 4.5 shows all the cases in which changes in the coefficients have been greater than ± 0.300. It is apparent that the vast majority of individual *decreases* have been concentrated in the two groups of social class and housing characteristics. Of the total of 128 such decreases, only 16 are not associated with either of these two groups. On the other hand, of the 54 cases of *increases* of over 0.300, 34 are found exclusively within the three remaining groups: the age structure, household tenure and household accommodation.

To summarize these somewhat complex patterns of change, it is evident that by including the council housing sector, not only do social class and the nature of housing become less diagnostic variables, but the age and house-hold composition variables become much more diagnostic. The council areas thus differ from the centrally located low-status private housing areas not only in that, while being of similar social class, they are better housed, but also in that they are very differently composed in terms of their age structures and the composition of their households. The actual changes of sign associated with the household variables, and the consistent increases in the values of coefficients associated with the age structure variables, illustrate

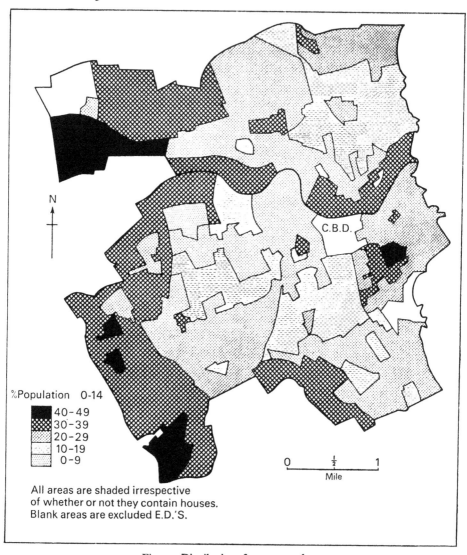

Fig. 4.3. Distribution of persons aged 0–14
Source : Census, e.d. data, 1961

that the social composition, as well as the housing characteristics of the council areas, are different in kind from the low-status areas in the private sector.

An interesting illustration of this is seen in the reduction of the coefficients associated with the percentage of females aged 20–24 who have ever married. The strong negative association between this variable and social class in the private matrix ($r = -0.472$) is reduced to a non-significant level ($r = -0.055$) in the matrix for the whole town. The negative association, which one might expect on the basis of literature on the subject, thus breaks down. The actual spatial distribution of the percentage of women married shows this to good effect. In the private sector, there is a well-defined pattern in which high incidence of young married women corresponds closely to areas of low social class, but by contrast the council areas, although low in social class, form a peripheral ring with low percentages of young women who are or have been married. The explanation lies in the fact that when young women from council areas do marry they are forced to seek accommodation in furnished or unfurnished privately rented housing since the local authority housing policy gives priority to families with large numbers of children, at the expense of childless couples. While the lowering of the association between low social class and high proportions of young females married does not therefore necessarily imply different behavioural patterns in the council housing areas, it does mean that in actual practice there is a markedly different pattern in the association between the variables in the private as against the council sector. To this extent, the council house areas are a radically different population.

Similar patterns of change can be seen particularly in the age structure and household size variables, as has already been noted. To a large extent, this is a reflection of the sorting-out process which operates through the local authority housing policy. The tendency to favour the larger families on the housing list, combined with the well-documented reluctance of older persons to move from central to peripheral areas (which frequently results in an exchange of houses where older people are offered council houses), both operate to give the council estates an ill-balanced population in terms of age structure. This has been one of the main conclusions from the comparison of the two matrices. It is also evident in the spatial distribution of the individual variables measuring age characteristics. Fig. 4.3, for example, shows how the distribution of persons aged 0–14 concentrates in distinct nodes of high proportions not only in the peripheral ring of council housing as a whole but also within particular estates and parts of estates within that ring. Such concentrations depend on the length of time that a particular estate has been occupied and, in consequence, upon whether the families within the area can be considered to be growing, static or in decline. The rather longer-established estates, such as Ford, Marley Potts and Hill View, for example,[1] have few children since most of the families have now reached maturity and many of the children who grew up on the estates have left the area or fall outside the range of ages in question. By contrast, the more recently built estates, such as Hylton Castle, Farringdon, Hylton Red House and parts of Penny-

[1] The locations of named council housing estates are shown in Fig. 3.10.

well, have very large proportions of younger children since the families are still at the expanding stage. In the most recently built estates such as Town End Farm there are low proportions of children aged 0–14 since the families are only just beginning to expand. Such variations in age structure have marked effects on the matrix of correlation for the whole town and have repercussions in a variety of social contexts. The Director of Education, for example, reported to the Education Committee in 1964 that, although the number of children reaching secondary school age from 1971 to 1974 would show little change in total from the number then in secondary schools, there would be a very marked change in the distribution between schools. At Hylton Red House the numbers would decrease from 175 to 120 between 1965 and 1974: at Hylton Castle they would decrease from 239 in 1965 to 157 in 1970 and to 87 in 1974: at Town End Farm, which it has been suggested has a large number of potentially expanding families, the numbers would increase from 97 to 211 between 1965 and 1970, but decrease to 187 by 1974.[1] Each estate thus tends to have a preponderance of people in a few age cohorts which act like waves creating bulges at successive age groups as the population ages. Such 'age bunching' has to be taken into account in analysing the town's social structure.

It was for this reason that two measures of fertility were included in the analysis. The most commonly used parameter is the number of children aged 0–4 as a percentage of females of child-bearing age (in this case taken as 15–44). This is an adequate measure where there is normal replacement of population and a roughly balanced age structure, but where, as in the council estates, there is high fertility but most of the families are at the same stage in the family cycle, the number of children aged 0–4 can understate considerably the fertility of a given area since large numbers of children just above this group are excluded. In an estate in which the age bulge has reached the age cohort 5–9, a fertility measure based on children aged 0–4 would obviously be a gross understatement of the area's fertility.

This can be shown diagrammatically, as in Fig. 4.4, where two measures of fertility have been plotted one against the other. Were the age structure of all the areas roughly balanced, one might expect a narrow scatter of points close to the regression line, and this indeed is apparent in most of the enumeration districts for which the two values have been plotted. Certain of the enumeration districts, however, show a marked deviation from the regression of one fertility ratio on the other, and it can be seen that such deviates are predominantly areas which lie in the three wards of Hylton Castle, Pennywell and Thorney Close, which correspond with the three

[1] Noted in *Population at Silksworth: report on population balance and house size*, Sunderland Corporation Planning Office Report, no. 2 (1965).

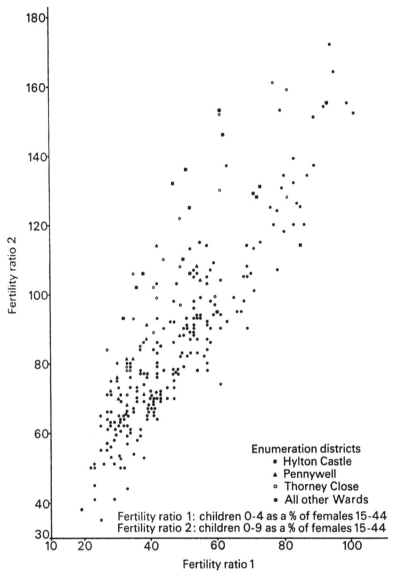

Fig. 4.4. Comparison of two alternative fertility ratios
Source : Census, e.d. data, 1961

council estates of the same names. All three are estates which are of fairly recent origin and in which the age-bulging has generally reached the age ranges above 0–4. It can be seen that most of these council estate enumeration districts have fertility ratios that are higher than expected when measured by the ratio of children aged 0–9 as against that for children aged 0–4. For example, an area in Hylton Castle with a ratio of 62, as measured by children aged 0–4, has a ratio of 154 as measured by children aged 0–9 whereas, judged by the majority of areas, it would be expected to have a ratio of about 100. To this extent the measure using children aged 0–9 might be considered to be a 'truer' estimate of the fertility of such council areas, and the other measure might be thought to understate the fertility.

The complications arising from these distorted age structures of the council estates produce some contradictory patterns in the correlation coefficients of the two fertility ratios. In both of the matrices, the two ratios show expected negative associations with high social class. Nevertheless between the two matrices the number of 'high' coefficients changes markedly between the two ratios. In the private matrix, the first ratio (using children 0–4) has no 'high' values whereas the second ratio (using children aged 0–9) has 14: in the matrix for the whole town the first ratio increases its score of 'high' values to a total of 4, whereas the second ratio decreases to 7. There are also some odd inconsistencies within the matrix for the whole town, produced by the distorted age structure of council estates. Within this matrix, for example, the first fertility ratio is negatively associated with the proportion of council tenants ($r = -0.342$) while the second ratio is positively associated ($r = 0.298$), and similar discrepancies between the two occur in their associations with 1-person, 2-person and 6-person households. While the previous discussion leads one to expect contradictions such as these, they nevertheless underline the diverse and complex effects of the inclusion of the council housing areas and spotlight, too, the complications of these two fertility measures in the context of council estates. Before more definite conclusions could be drawn more detailed knowledge and field work would be necessary.

It can therefore be seen that in addition to making the classical type of ecological analysis largely invalid, the council house areas introduce new elements and new combinations of traits into the structure of this town and, doubtless, of other towns. They are areas in which low class is no longer associated with poor housing (even though the association with high density per room persists), and in which, in terms of age structure and household composition, the population differs radically from areas within the private housing sector regardless of whether such areas are of high or low social class.

MULTIVARIATE ANALYSIS

It is because of such disruptions caused by council housing and because of the greater degree of town-planning control and of social and geographical mobility within the present-day social structure of Britain as compared with the nineteenth or even the early twentieth centuries, that a new approach to urban analysis is needed. It is because of the complexity of trying to handle a large number of interrelated variables in the form of simple associations such as have been dealt with above, that some form of multivariate technique is appropriate. The reasons underlying the selection of principal components analysis have already been discussed in an earlier chapter, where it was seen that the value of the technique was its ability to isolate orthogonal bundles of interrelated phenomena in which the variables or parts of variables measure aspects of the same overall cluster. The discussion of the simple associations of the correlation matrices has suggested that certain variables are more diagnostic indexes than others. If one were to isolate clusters of interrelated variables one of these clusters might be expected to be related to social class, which in various guises can be held to account for a large number of the simple associations. Another might be expected to relate to the physical nature of the housing; a third might be expected to group together aspects of the age structure and so forth. Indeed, the very rough grouping of the 30 variables was a rather clumsy attempt to isolate such clusters. Ideally the object of the component analysis is therefore to carry the analysis a stage further by isolating such clusters in a more precise, quantifiable and objective fashion.

Before doing this, however, a number of cautionary points must be made about the matrices of correlation coefficients on which the previous discussion has been based and from which the component analysis is derived. Three main warnings seem most pertinent to any interpretation of results based upon the correlation matrices.

Ecological correlations

In making correlations between sets of data grouped on an areal basis, the significance and limitations of the associations must be kept clearly in mind. As with aggregation in general, the use of group data rather than individual data introduces difficulties of interpretation. Early work by ecologists tended to infer, either explicitly or by implication, that one could argue from ecological, that is group, correlations through to individual correlations. For example, the existence of a positive association between high crime rates and large numbers of, say, Japanese, in an area might lead one to suppose that, at an individual level, Japanese are criminal. Yet, especially since both would tend to have low absolute rates of incidence in a Western population, there is no reason to suppose that individual Japanese in such an area would have higher rates of crime than non-Japanese. They may be a law-abiding minority which happens to live in areas characterized by high rates of crime. The ecological correlation, in other words, applies to the area and not necessarily to the individuals within the area. Robinson has made the point most forcibly. He argues that, given a fourfold table summarizing the frequencies of two attributes, the individual correlation depends upon the internal frequencies of the within-areas individual correlation, while the ecological correlation depends upon the marginal frequencies. Since there is a large number of combinations of internal frequencies which will each produce the same marginal frequencies, there is no necessary correspondence between individual and ecological correlations.[1] From the resulting body of literature which has arisen, it has been pointed out that Robinson somewhat overstated his case. Duncan and Davis, for example, have shown that in certain circumstances areal relationships could be used to estimate individual associations.[2] More important however is the fact that much ecological research has been more interested in areal associations than in individual associations. In such cases, the use of areal data is not a second-best substitute for data at an individual level.[3] This is especially the case where, as in the present study, the characteristics which are used are not minority traits. Were the centre of interest criminal tendencies or the incidence of mental disease, the data would have to be taken to an individual level in trying to provide explanations. As it is, it is sufficient to note the difficulty and to bear in mind that the correlations, and the later analysis based upon these correlations, refer not to individuals but to the enumeration districts in which the individual lives.

Size of areal unit

In considering the size of the areal units used, two distinct problems are presented: first the question of variation in the population size of the enumeration districts, and second the difficulty of using data drawn from both the 100 per cent and 10 per cent data of the 1961 census.[4]

[1] W. S. Robinson, 'Ecological correlation and the behaviour of individuals', *Am. Sociol. Rev.* xv (1950), 351–7.

[2] O. D. Duncan and B. Davis, 'An alternative to ecological correlation', *Am. Sociol. Rev.* xviii (1953), 665–6. For further reactions to Robinson's paper see L. A. Goodman, 'Ecological regressions and the behaviour of individuals', *Am. Sociol. Rev.* xviii (1953), 663–4, and *idem*, 'Some alternatives to ecological correlation', *Am. J. Sociol.* lxiv (1959), 610–25. For an attempt to resolve conflicting views see O. D. Duncan, R. P. Cuzzort and B. Duncan, *Statistical geography: problems in analyzing areal data* (Glencoe, Ill., 1961), pp. 65–7.

[3] D. J. Bogue, 'Population distribution', in Hauser and Duncan (eds.), *The study of population*, p. 392.

[4] The 1961 census of population was collected at two levels of coverage: most of the data were gathered for every individual (the 100 per cent data); certain questions, including those relating to occupation and education, were asked of a 10 per cent sample of the total population (the 10 per cent data). The sample was drawn simply by submitting the longer questionnaire to every tenth household.

The first difficulty again relates to the aggregation problem, in this case to the question of scale or to the problem of 'areal levels of generalization'. As early as 1934, Gelkhe and Biehl pointed out that the size of the areal units used in correlation affects the size of the correlations obtained.[1] Subsequently a number of studies have shown that the use of different areas can alter not only the size, but also the direction of association of the variables being considered. Duncan, Cuzzort and Duncan, for example, have shown how the population of the United States appears to be growing more concentrated over time if county data are used, yet more dispersed using data for states.[2] One of a number of suggestions for the standardization of areal units is Robinson's argument for the use of a weighting factor based on the areas of each unit of observation.[3] While it has been pointed out that Robinson's formulae apply only under two special conditions—either when the values for the aggregated areas are equal or when such changes through aggregation are strictly proportional[4]—the principle of weighting for unequal units appears sound. In the case of the Sunderland data, for example, the important criterion would seem to be population size rather than area. In the 100 per cent data the range of size is from 75 persons to 1,313 persons; in the 10 per cent data the corresponding range is from 4 to 157. The median values are 630 and 65 respectively and the means and standard deviations are shown in Fig. 4.5. While the range of population size is not too wide, it was decided, as a first step in overcoming the difficulty of this variation, to ignore all those enumeration districts which had populations of less than 30 in the 10 per cent data and less than 300 in the 100 per cent data. These figures mark distinct breaks in the frequency distributions of Fig. 4.5 and correspond roughly with the lines drawn at two standard deviations from the mean. Effectively it excludes the most aberrant areas, those for which averages are most suspect. In all, nine enumeration districts were thus excluded, five of which fell within the area designated as the Central Business District while the remaining four were either peripheral undeveloped areas or areas in the east of the town which were composed of industry and decayed or partly demolished housing.[5] The original 272 enumeration districts delimited by the census were therefore reduced to 263 with a population range of 330 to

[1] C. E. Gelkhe and K. Biehl, 'Certain effects of grouping upon the size of the correlation coefficient in census tract material', *J. Am. statist. Ass.* XXIX (1934), 169–70.

[2] Duncan, Cuzzort and Duncan, *Statistical geography*, pp. 82–90.

[3] A. H. Robinson, 'The necessity of weighting values in correlation analysis of areal data', *Ann. Ass. Am. Geogr.* XLVI (1956), 233–6.

[4] E. N. Thomas and D. L. Anderson, 'Additional comments on weighting values in correlation analysis of areal data', *Ann. Ass. Am. Geogr.* LV (1965), 492–505. For further discussion of Robinson's suggestions see Haggett, *Locational analysis in human geography*, pp. 205–10.

[5] The nine excluded e.d.s are shown blank in the figures accompanying the discussion of the component analysis below. The nine e.d.s are listed on p. 266.

1,313. To allow for this remaining size variation a correcting factor based on the square root of population size was incorporated into the calculation of correlation coefficients.[1]

The second difficulty, arising from the use of 10 per cent data, was more intractable. The 10 per cent data of the 1961 census cover most of the social

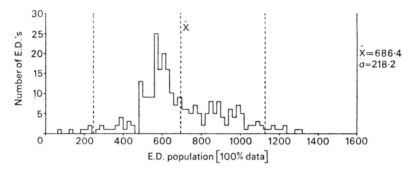

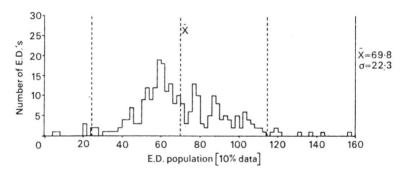

Dashed lines are shown at two standard deviations from the mean in both populations

Fig. 4.5. Frequency distributions of population sizes of enumeration districts (E.D.) in Sunderland—100 per cent and 10 per cent data

class variables whose exclusion from the analysis would have made an unfortunate omission. Yet the discovery of serious bias in the sample data has thrown doubt on the validity of the measures.[2] Correcting factors have been calculated by the General Register Office, but only for regions within

---

[1] Since the standard error can be taken as being roughly equal to the square root of size, the weighting was based on the square root of the population of each e.d. ($\sqrt{P}$). The appropriate formula thus becomes
$$r = \frac{\Sigma\sqrt{P}.\,\Sigma\sqrt{P}xy-(\Sigma\sqrt{P}x)\,(\Sigma\sqrt{P}y)}{\sqrt{\{[\Sigma\sqrt{P}.\,\Sigma\sqrt{P}x^2-(\Sigma\sqrt{P}x)^2].[\Sigma\sqrt{P}.\,\Sigma\sqrt{P}y^2-(\Sigma\sqrt{P}y)^2]\}}}$$

[2] Bias in the 10 per cent sample data was first reported in a private communication from the General Register Office on 23 November 1964. Recent census reports list regional and other correcting factors.

the country or for groups of urban areas. At the level of the enumeration districts, the task of calculating such factors is virtually impossible and to apply regional correcting factors would appear to be likely to introduce as many errors as it might perhaps correct. Indeed, opinion on the advisability of using 10 per cent data in urban analysis seems to be equally divided. On the one hand, Gittus, for example, has excluded 10 per cent data from her analyses of Hampshire and Merseyside, arguing that the calculation of measures of significance for biased samples is statistically meaningless. The Centre for Urban Studies, on the other hand, has included 10 per cent data in its analysis of London, trusting that the sampling errors will come out in the statistical wash.[1] In incorporating data from the 10 per cent sample, one can merely note that local knowledge does not suggest any major inaccuracies in the data and that its exclusion would have severely emasculated much of the analysis.

Non-linearity of the distributions

Product-moment correlation is a measure of linear association. Many of the distributions of the social and economic data of the 30 variables were clearly not linear. To this extent, the coefficient will tend to understate the amount of association which exists between sets of data. One method of allowing for such non-linearity would have been the use of Spearman's rank correlation coefficient. Alternatively the data might have been transformed to logarithmic values. The fact that Moser and Scott found that neither method made significant differences to the results of their correlations in a similar exercise, lends some justification to ignoring the non-linearity of some of the data.[2]

Such warnings have to be borne in mind in considering the conclusions which have already been drawn concerning the correlation matrices and also in considering the results of the component analysis which follows.

COMPONENT ANALYSIS

Component analysis does not rely on any underlying premises which predetermine the data selected. However, while it is not trammelled by the Procrustean bed of possibly suspect theory the data which are fed into the process do determine the nature of the components which emerge. Since the components are merely linear combinations of the original data, the selection of the initial variables is of the utmost importance. In the present study, all

[1] Gittus, 'An experiment in the definition of urban sub-areas', *Trans. Bartlett Soc.* XI (1964–5), 109–35; Centre for Urban Studies, 'A note on the principal components analysis...for London'.
[2] Moser and Scott, *British towns*, pp. 56–60.

the 30 variables were selected with an eye to their possible theoretical import and the aim was to select a sufficient cross-section of variables so as not to give undue weight to any one aspect of urban social structure (see Table 4.1). It must be borne in mind, however, that the following analysis must be seen simply as a manipulation of the particular variables which compose the universe of data. A second consideration is that the analysis can only be relative to the conditions found within the set of data being used, since the output is dependent entirely on the input. It is only internal differences and likenesses within the set of data that is used which can be isolated. It was partly for this reason that Sunderland was compared with other British towns in an earlier chapter, so that it would be set within the national context.

The analysis of the data begins from the correlation matrix for the whole town and the components are derived from the latent vectors (l_{1-30}) of the matrix.[1] The latent root (λ) of each component is the sum of the scores of the variances of each vector. Since the correlation coefficients are a means of standardizing the variables with unit variance, the sum of individual variance for each variable is 30 and the latent root of each component therefore shows the proportion of the total variance which is accounted for by that component. Table 4.7 shows the results of this analysis for the 30 variables over the whole town. Extracted in order of importance, only the first five components are shown and their latent roots are 9·07, 8·60, 2·37, 1·85 and 1·13. In all, therefore, 23·02 of the total variance (30) has been accounted for. In other words over three-quarters (76·75 per cent) of the variation in the total universe has been 'explained' by these five components. It has been noted earlier that this proportion is similar to those which have been found in comparable studies of census material in urban areas.[2]

Table 4.7 also shows the elements appropriate to the latent vectors of each of the 30 variables. These vectors were initially extracted so that the sum of their squares was equal to unity, but by weighting these values the vectors have been expressed so that the sum of their squares equals the latent root of each component in turn. The figures are therefore equal to the correlation coefficients between each variable and the component in question and they can therefore be used to analyse the composition of each component to isolate precisely what each is measuring. It must be remembered, of course, that in trying to interpret the components, as Kendall says, 'in many cases our principal components do not have an identifiable separate existence

[1] The standard work on multivariate analysis is H. H. Harman, *Modern factor analysis* (Chicago, 1960). Texts slanted to the analysis of social data are W. W. Cooley and P. R. Lohnes, *Multivariate procedures for the behavioral sciences* (New York, 1962), and R. B. Cattell, *Factor analysis* (New York, 1952). For an excellent, lucid and short text see M. G. Kendall, *A course in multivariate analysis* (London, 1957), which is the work principally used here.
[2] See above, pp. 64–7.

TABLE 4.7. *Latent vectors of components 1–5*

Variable	\multicolumn{5}{c}{Components}	Percentage variability absorbed by components 1–5	Percentage of total variability accounted for by each variable
	1	2	3	4	5		
1	0·78	−0·13	−0·35	−0·10	−0·00	76	12
2	0·78	−0·08	−0·12	−0·01	−0·14	65	11
3	−0·51	−0·01	0·38	−0·18	0·30	53	9
4	−0·72	0·18	−0·00	0·25	−0·20	65	17
5	0·91	−0·15	−0·23	−0·14	0·03	92	26
6	−0·84	0·13	0·35	−0·04	−0·10	86	23
7	−0·07	−0·25	0·16	0·52	−0·18	40	4
8	0·00	0·66	−0·09	−0·49	0·06	69	14
9	0·21	−0·72	−0·37	−0·24	−0·04	77	18
10	−0·75	−0·32	−0·42	−0·33	0·04	95	22
11	−0·01	0·54	−0·12	0·53	0·28	67	10
12	0·64	0·49	0·35	0·13	−0·03	80	21
13	−0·14	0·59	−0·12	−0·18	0·48	65	12
14	−0·64	0·08	−0·31	−0·54	0·04	80	15
15	0·88	0·05	0·09	−0·24	−0·05	85	24
16	−0·60	−0·68	−0·07	0·17	0·01	86	24
17	0·21	0·58	−0·39	0·29	0·42	79	13
18	−0·10	0·90	0·04	−0·02	−0·04	82	23
19	0·30	0·76	0·16	0·12	0·04	72	20
20	0·72	0·46	0·31	−0·05	−0·00	83	23
21	−0·65	−0·45	−0·19	0·33	0·06	78	20
22	−0·92	−0·01	−0·13	−0·02	0·03	86	26
23	−0·68	0·40	−0·15	0·04	0·06	65	19
24	0·60	−0·68	−0·28	0·03	0·08	91	25
25	0·29	0·37	−0·60	0·33	0·19	72	10
26	0·24	−0·77	0·22	−0·01	0·05	70	19
27	0·09	−0·64	0·46	−0·05	0·47	86	15
28	−0·05	0·71	−0·46	−0·07	−0·38	87	17
29	−0·19	0·87	0·15	−0·19	−0·11	86	23
30	0·08	−0·93	−0·08	0·11	0·00	90	25
Latent root (λ)	9·07	8·60	2·37	1·85	1·13		
Percentage of total variability accounted for by each component	30	29	8	6	4		

NOTES. *i* Elements appropriate to the latent vectors are expressed so that, for each component, the sum of squares of the vector elements equals the latent root of that component, i.e. $\Sigma l^2 = \lambda$.
ii Variables are listed in Table 4.1, p. 136.

and are to be regarded as convenient mathematical artefacts'.[1] In some cases, and as an instance Kendall cites the 'general intelligence' factor, it is arguable whether the components can be given any reality. Nevertheless, if one examines the variables with which each of the Sunderland components is most closely associated, one can glean some fairly clear impressions of the general 'mix' which comprises each of the principal components.

Component 1

Original variables with the ten highest associations with component 1

		Vector elements
22	Average number of persons per room	−0·920
5	Social class score	0·907
15	Percentage owner occupiers	0·882
6	Percentage with terminal education age below 16	−0·840
1	Percentage social class I	0·780
2	Percentage social class II	0·780
10	Percentage aged 0–14	−0·747
4	Percentage social class IV	−0·720
20	Percentage 2-person households	0·720
23	Percentage households at over 1½ persons per room	−0·678

In each case the signs of the elements appropriate to the vectors indicate the direction of the association while the size of each value indicates the degree of association of each variable with the component in question.

The list of variables most highly correlated with component 1 clearly shows that it is closely associated with a range of social class indexes. It is most highly associated, negatively, with the average number of persons per room which alone accounts for as much as 84·6 per cent of this first component. The second highest association is with the compound social class score, and, among the ten highest associations, there are three other direct social class parameters all of which show that the component is positively associated with high social class. One interesting feature is the fact that the percentage of persons aged 0–14 is the seventh most highly associated variable, with a negative association of −0·747. This is a reflection of the lower fertility ratios of the higher rated areas and this, indeed, is supported by the fact that the second fertility ratio also has a high negative association with component 1. In view of the use of rateable values in the last chapter, it is also interesting to note that, while there is a moderately high positive association between gross rateable values and component 1, the value is only 15th highest out of the total of 30 variables. Recalling that the analysis is derived from the matrix for the whole town, this again underlines

[1] Kendall, *Course in multivariate analysis*, pp. 26–7.

the lesser usefulness of rating values when the council sector is included in the analysis.

The associations of component 1 are therefore clearly delineated, and show that it is positively associated with high social class.

Component 2

Original variables with the ten highest associations with component 2

		Vector elements
30	Percentage households with all four amenities	−0·935
18	Percentage renting private unfurnished accommodation	0·900
29	Percentage without exclusive use of fixed bath	0·868
26	Percentage un-subdivided houses	−0·771
19	Percentage 1-person households	0·762
9	*J*-index	−0·721
28	Percentage shared dwellings	0·707
24	Gross rateable value	−0·683
16	Percentage council tenants	−0·677
8	Percentage females aged 20–24 ever married	0·663

Component 2 is likewise relatively easily identifiable. The individual associations show that it is a measure of housing conditions. The association with the percentage of houses with all four amenities is highly negative and, alone, explains 87·4 per cent of the component. The high positive association with the privately rented unfurnished variable shows that poor housing conditions are highly concentrated in the privately rented housing sector (the privately rented furnished index also shows a high positive correlation with component 2: the fact that it is less high than the unfurnished index is probably due to the much smaller proportion of houses which are rented furnished). The other high associations are consistent with an interpretation of this component as a measure of poor housing: for example, the associations with subdivided housing and shared accommodation are both highly positive. An interesting feature is the high positive association with the proportion of people aged 65 and over, where the latent vector is 0·490. This picks up the age-segregating effects of council housing which has tended to concentrate the older-aged 1- or 2-person households in the privately rented sectors of the town where housing conditions are poor. The vector elements all point to such patterns: there is a high negative association with council housing, a high positive association with 1-person households and with persons aged 65 and over.

In parenthesis, it is also interesting to note that rateable value has a high negative association with component 2. Indeed Table 4.7 shows that it is relatively highly associated with each of the first three principal components. In none is the association very great. To this extent, rateable value fails to differentiate between the three factors, or clusters, which the component analysis identifies. It would therefore appear to be a blanket measure which combines aspects of a large number of different items, but cannot discriminate between them.

Component 2 thus emerges clearly as being positively associated with poor housing conditions and with the age and household traits associated with poor housing. Its negative association with the council house sector is therefore an expected consequence.

These two principal components, isolating the social class and housing elements in the matrix, together account for as much as 58·9 per cent of the total variability in the matrix of associations and their respective contributions to this explanation are approximately equal. If one were to consider only these two components, nearly two-thirds of the total data would have been reduced to two composite measures each of which has clearly delineated characteristics. Turning to the next highest component, there is a large drop in the amount of variability which is accounted for.

Component 3

Original variables with the ten highest associations with component 3

		Vector elements
25	Percentage dwellings with over 8 rooms	−0·599
28	Percentage shared dwellings	−0·465
27	Percentage households with 1 household space	0·459
10	Percentage aged 0–14	−0·419
17	Percentage renting private furnished accommodation	−0·389
3	Percentage social class III	0·383
9	*J*-index	−0·374
1	Percentage social class I	−0·355
12	Percentage aged 65 and over	0·351
6	Percentage terminal education age below 16	0·349

Component 3 'explains' only 7·9 per cent of the total variability and this is reflected in the appreciably lower vectors of the variables. Nevertheless, these weightings do make a coherent pattern which, while less easily identified than the first two components, can be defined. The negative associations with large houses, shared dwellings, and the privately rented furnished housing sector all point to a consistent negative association with the factors of 'invasion' and subdivision which were stressed in the last chapter. Essentially, with the exception of the negative association with children (aged 0–14), this third component appears to be isolating, negatively, those areas of subdivided housing which have suffered loss of value over time. The positive association with the proportion of households with one household space adds emphasis to this.

There is also a subsidiary, although much less important, syndrome running through the composition of this component since the vectors show a consistent, albeit not very strong, association with low social class. This is seen in the negative associations with Social Class I and the *J*-index, and in the positive association with those with low school-leaving age.

The only remaining feature of the list of highly related variables which calls for comment is the negative association with persons aged 0–14. Were component 3 simply a negative measure of the classic rooming-house type of area, one would expect this association with children to be positive since such areas typically contain few children. The fact that subdivision of housing and large proportions of children are associated in the same direction with component 3 suggests that no distinction is being drawn between those two types of subdivided area which were isolated in the previous chapter. This point will emerge more clearly in looking at the pattern of the fourth component, where such a distinction, based on age structure, is drawn.

Component 4

Original variables with the ten highest associations with component 4

		Vector elements
14	Fertility ratio 2 (children 0–9)	− 0·537
11	Percentage aged 20–24	0·523
7	Percentage females in labour force	0·522
8	Percentage females aged 20–24 ever married	− 0·488
10	Percentage aged 0–14	− 0·332
25	Percentage dwellings with over 8 rooms	0·330
21	Percentage 6-person households	0·328
17	Percentage renting private furnished accommodation	0·287
4	Percentage social class IV	0·252
9	*J*-index	− 0·244

This combination of variables in component 4 is a most intriguing mixture since it groups together many items which are somewhat similar to Shevky's index of *urbanization*.[1] There is a negative association with fertility, as measured both by fertility ratio 2 and by the proportion of persons aged 0–14; there is a positive association with the proportion of females in the labour force and with indexes of subdivided housing (in so far as this is partly related to large houses and the privately rented furnished housing sector). There are also associations with other variables not considered by Shevky, such as the proportion of females aged 20–24 who have ever married, and the age category 20–24.

One difference between this measure and Shevky's index is that the combination of items included here as component 4 has emerged from the statistical handling of the data rather than from theory. Each item has been given a

[1] Shevky and Bell, *Social area analysis*. For discussion, see above, chapter 2.

weighting appropriate to its respective importance. Furthermore, while the composition of component 4 may appear to lend strong support to the Shevky analysis, if one examines the associations between the relevant variables more closely, it appears that the support for the Shevky index of *urbanization* is illusory. The measures of fertility and women in the labour force can be shown to be related not to the concept of 'urbanism' but rather to poverty or economic want. This can be made clearer by looking at the spatial distribution of the raw values of the two variables. It is noticeable, for example, that many council estate areas have high proportions of women working, but there is a marked difference between those estates which are of a few years' standing and those which are more recent. The rather older estates, in which the majority of families are beyond the infant stage, have consistently high proportions of women in the labour force, whereas the newer estates have lower proportions. This contrast can be seen between the pre-1950 Hill View estate and the recently built Hylton Castle estate. It would therefore seem that the association between fertility and women at work is not an indication of a poorly developed family life, as Shevky would argue, but rather that it reflects a state of affairs in which families have reached that stage in their growth cycle at which, with children being old enough not to need such close attention, mothers are freed for work. All told, this association would seem to be more a measure of poverty than of so-called *urbanization*. With a need for more household income, in low-income-group families, the desire for females to work is always present, but whether they work or not is controlled in practice by the age of the children within the family. Like the water through a plumbing system, the pressure is provided by poverty, while the actual flow is determined by the tap of the age structure.

This component is therefore best considered as being associated with a number of social characteristics associated with the age structure, and with the housing characteristics associated with subdivision. To this extent, it is partly a mirror image of component 3 but, unlike component 3, it adds an age structure differential to its isolation of social characteristics.

Component 5

Component 5 is the most difficult one to which to give a specific identity. This is not surprising in that one is dealing with a component which 'explains' as little as 3·8 per cent of the total variation so that the total proportion of the variation of each variable which is accounted for is correspondingly low. As one moves to successively lower components they become increasingly difficult to interpret. The elements appropriate to each vector in Table 4.7 show that the variable most highly related to component 5 is the fertility ratio (using children aged 0–4) with a vector weight of 0·483: the second most highly related is the proportion of households with one household space, with a weight of 0·474. The association with the remaining variables falls rapidly and includes some apparently contradictory patterns. For example, both the percentage of households with one household space and the percentage of privately rented furnished accommodation are weighted positively, which is an unexpected combination. Such difficulty of interpretation is not un-expected in view of the small proportion of variation which is accounted for by component 5. No doubt it, and the subsequent components, are largely absorbing the remainder of the variability in combinations which are statis-tically significant but which, in terms of the chosen variables, are impossible to identify. The remainder of the analysis will therefore be restricted to the first four components, which are more readily identified and which together account for nearly three-quarters of the total variation in the universe of input data.[1]

COMPONENT SCORES

Having identified the content of each of the first four principal components, it is now possible to allot scores appropriate to each of these components for each of the 263 enumeration districts which were used in the analysis. This is done by using the elements of the latent vectors of each variable, for each of the four components in turn, as weights which are applied to the original data expressed in standardized form. Thus for each enumeration district the score for component 1 is calculated as $l_1 z_1 + l_2 z_2 \ldots l_{30} z_{30}$, where l_i is the element appropriate to the ith variable and z_i is the standardized raw value of the ith variable for the enumeration district in question. This can be expressed in general terms as

$$\sum_{i=1}^{30} l_i \left(\frac{x_i - \bar{x}_i}{\sigma_i} \right)$$

where x_i, $\bar{x}_i$ and σ_i are the raw value, the mean and the standard deviation of the ith variable. The details of the calculation of these scores and a suggested short-cut method are discussed in Appendix C. The four sets of component scores which were produced for the 263 enumeration districts have unit variance and zero mean and the histograms of their overall distribu-tion for each component are shown in Fig. 4.6. These frequency distributions

---

[1] It is interesting to note that, if Weaver's combination indexes are applied to the latent roots, the minimum sum of squared deviations from the ideal distribution is found to correspond with the distribution with four components. This suggests further evidence in support of the use of only the first four components (i.e. excluding component 5 and beyond). The respective sums of squares are as follows: for one component 5,811; for two 961; for three 729; for four 701; for five 779. Cf. J. C. Weaver, 'Crop combination regions in the Middle West', *Geogr. Rev.* XLIV (1954), 175–200.

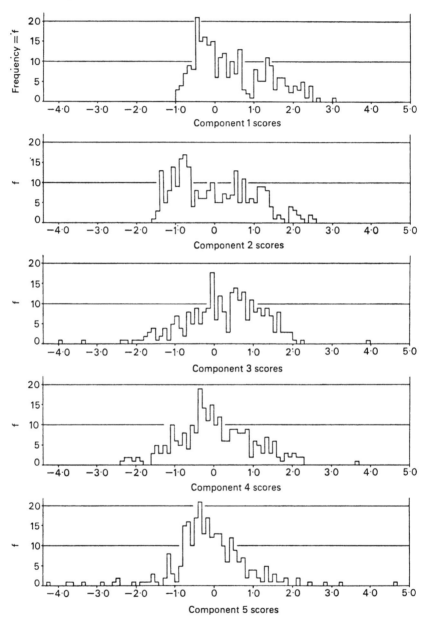

Each distribution is standardised to zero mean and unit variance

Fig. 4.6. Frequency distribution of component scores

provide a guide to the selection of cutting points for the mapping of the component scores. Since the skewness of the distributions varies, standard cutting points for all four sets of scores would seem to have disadvantages and, instead, in showing simplified maps of their distributions, various cutting points have been used based on standard deviations. The actual scores for each enumeration district (e.d.) are listed in Appendix D. The figures which accompany the text show only a simplified distribution of these scores. In discussing the patterns which emerge, both these maps and the detailed scores in Appendix D will be referred to.

Component 1 scores

Recalling the designation of component 1 as a social class component, the pattern of its scores (Fig. 4.7), not unexpectedly, closely resembles the by now familiar pattern produced by rating values and the distributions of direct social class measures. Being most highly associated, negatively, with the average number of persons per room, it is also a mirror image of that distribution.

There are two main areas of high scores: a wedge of high values in the south of the town in which the highest values are found in the middle class area of St Michael's Ward e.d.s 25 and 26;[1] and a second area in the north with its highest values adjacent to the sea coast in Fulwell e.d. 45 and Roker e.d. 35. In contrast to these areas, the areas adjacent to the river and those close to the sea to the south of the river mouth have consistently low scores. So too has the girdle of council estates in the west of the town. The pattern which emerges is therefore a very accurate reflection of all that has so far been said as regards the social class structure of the town. As a single composite score the distribution shown in Fig. 4.7 tells one a great deal about the various types of social areas found within Sunderland. Indeed, as has been shown, the component subsumes 30 per cent of all the variability of the initial 30 variables.

The scores of individual enumeration districts also accord very closely with the details of the social class structure of particular areas. For example, the council areas built in the inner parts of the town are well differentiated from surrounding areas of private housing. In Bishopwearmouth, this replacement of old decaying rooming-house areas by council housing is reflected in the lower scores of Bishopwearmouth e.d.s 33 and 34. Similar council develop-

---

[1] In referring to the detailed pattern of scores listed in Appendix D, the following convention is used. Individual enumeration districts are referred to first by the name of the ward in which they lie and then by the reference number of the enumeration district. Thus St Michael's e.d. 25 is enumeration district 25 in St Michael's Ward. The location of the e.d.s is shown in the map in Appendix D, p. 267.

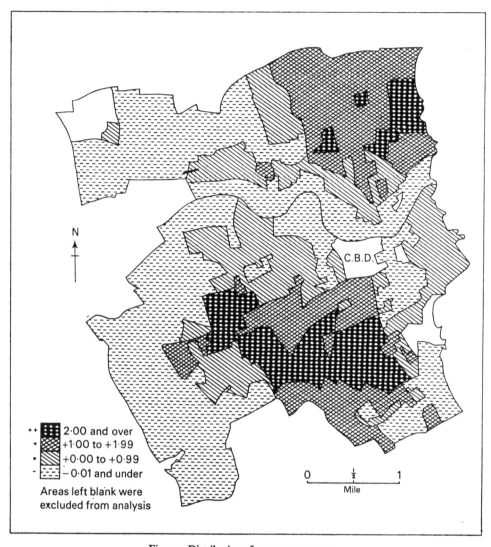

Fig. 4.7. Distribution of component 1 scores

ments can be seen in the lower scores of Bridge e.d. 41 and in Roker e.d. 34, which are areas of inner council housing. In the north of the town, the old village nucleus of Fulwell, which houses a rather different population from that of the surrounding newer residential areas, is differentiated by the lower scores of Fulwell e.d. 40. Finally, mention might be made of the recently built private housing estate of Seaburn Dene in the north of the town (Fulwell e.d.s 47 to 50 inclusive). These areas have markedly lower scores on component 1 than do other private middle-class residential areas, which is a reflection of the changing social differentials and the spread of lower middle- and upper working-class people into new private housing estates. With the wider extent of owner occupancy and the building of cheaper private residential estates, this, and some similar new estates, tend to include a much more mixed population in terms of social class. It is worthy of note that component 1 picks up such differences.

In a few enumeration districts, where the boundaries include very different types of area, an average score results which measures neither one nor the other part of such composite areas. Such is the case in St Michael's e.d. 18, in Humbledon e.d. 63 and in Colliery e.d. 54. In these areas, middle-class private housing and council housing development have unfortunately been included within a single enumeration district and the resulting scores are largely meaningless, since they conceal the intra-enumeration district heterogeneity which such boundaries produce. With such exceptions, however, the pattern of scores of component 1 is an accurate and discriminating single measure of a variety of social-class aspects of the town.

Component 2 scores

Component 2 is a measure of poor housing conditions, and the scores reflect this in their essentially concentric pattern: the zone of high scores around the central areas, indicating poor housing conditions, falls with a fairly regular gradient as one moves outwards to the lower scores of the better housing of the peripheral areas. This pattern roughly follows the historical growth of the built-up area and, in fact, the division between the third and fourth sextiles of values occurs at roughly the division between nineteenth- and twentieth-century housing areas, with higher scores being found in the nineteenth-century areas. The only nineteenth-century areas which do not have high scores are the areas of large undecayed housing in Thornhill and St Michael's Ward.

The fact that housing amenities are the most important single measure in component 2 means that the scores cut across the pattern of scores on component 1 and hence link together the council housing areas and the private middle-class areas in the north and south of the town. Indeed, with the exception of the private middle class areas of St Michael's (e.d.s 21, 22, 23, 25 and 28) and one area in Humbledon (e.d. 69), all the areas falling within the lowest sextile of values are found within council house areas. This pattern is seen, in a generalized fashion, in Fig. 4.8. Within the low-scoring areas, it is interesting to note that the individual e.d. scores accurately differentiate between the peripheral council estates and the inner redevelopment areas.

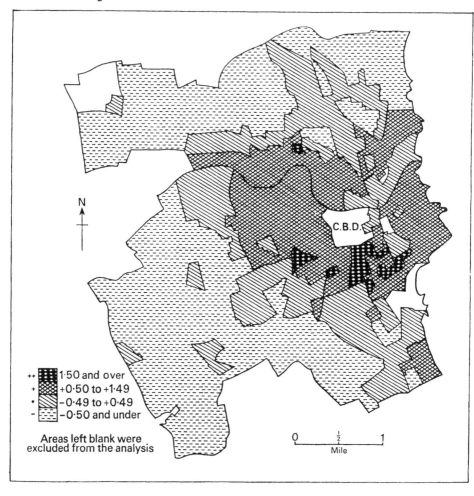

Fig. 4.8. Distribution of component 2 scores

The inner areas do not have such low scores since most of the development is in the form of high density flats, which serves to differentiate the two types of council property in terms of their housing characteristics. Nevertheless, such inner council house areas are clearly separated from the surrounding private housing areas. Bridge, e.d. 41, Central e.d.s 24, 30 and 31, and Monkwearmouth e.d. 10, which are areas of centrally built local authority redevelopment schemes, all have noticeably lower scores than the surrounding private areas. Furthermore, within the peripheral ring of council estates, a differentiation emerges between the newer and the older areas. The early-built parts of Ford estate (covering most of the north part of Pallion Ward),

Marley Potts (Southwick e.d.s 5, 6 and parts of 3 and 7), Newcastle Road (Colliery e.d. 55), Hill View (St Michael's e.d.s 16, 17 and parts of 14 and 18), and Plains Farm (Humbledon e.d.s 56 to 61 inclusive), for example, which are relatively early council areas, all score more highly than the more recent waves of council estate areas whose lower scores reflect the generally better housing standards.

Finally, within the private housing area, one might note that the old nucleus of Fulwell is again differentiated in terms of component 2. Fulwell e.d. 40 and Roker e.d. 33 both have higher scores than the surrounding areas, suggesting their relative poverty in housing amenities.

The pattern of scores for component 2 would therefore appear to be consistent with what is known of the town's housing characteristics and to add a further areal differentiation based on valid criteria.

Component 3 scores

The distribution of component 3 scores, shown in Fig. 4.9, links together an interesting combination of areas. There is a wedge of low scores in the south of the town embracing the rooming-house district of Bishopwearmouth, the subdivided low-class area of Hendon, and the middle-class area of St Michael's. In the north, low scores are found in a belt running northwards from Monkwearmouth adjacent to the sea and spreading west at the northern boundary to include the new private housing estate of Seaburn Dene. Low scores are also found in the extreme west of the town in the council housing estates. By contrast, high scores are found in the low-class private housing areas (excluding all the highly subdivided areas in Hendon and Monkwearmouth), and in the older generation of council estates.

Recalling the composition of this component, it seems necessary to take two factors into account in understanding this distribution. First is the subdivision of housing. This accounts for the very low scores in the two rooming-house areas of Bishopwearmouth (most of Bishopwearmouth Ward and St Michael's e.d.s 19, 20 and 24) and Roker (Roker e.d.s 27 and 28), and in the low-class subdivided areas of Hendon (Park Ward) and Monkwearmouth (Monkwearmouth e.d.s 11 to 17 and 20 and 21). These are areas in which, as has been shown, large houses have become subdivided and, in the rooming-house areas, a large proportion of the tenancies are in furnished accommodation. The second factor is related to the age characteristics of the areas. The predominance of elderly persons in the private housing low-class areas accounts for the isolation of these areas as being of high scores on component 3. The age structure connections of component 3 also explain the different scores which separate the newer and older council estates. In component 3,

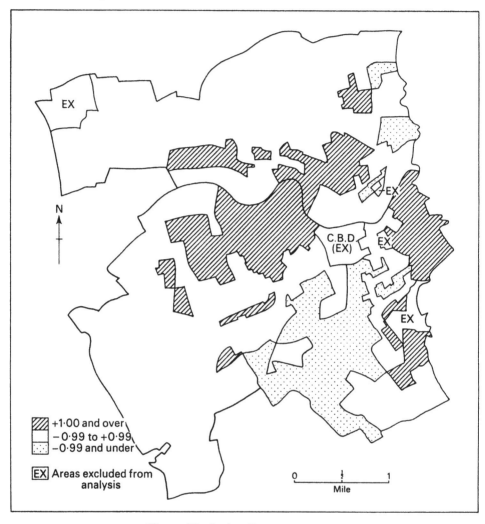

Fig. 4.9. Distribution of component 3 scores

this differentiation is greater than it was in the scores of the second component. The older estates have scores which fall in the highest or second highest sextile of values whereas the newer estates are almost invariably well below this. One exception is the council estate of Town End Farm (Hylton Castle e.d. 28) which is the most recent of the peripheral estates. The high score found here is doubtless a consequence of the preponderance of families which fall within the 'growing' stage of the family cycle.

The low scores which are found in the private housing middle-class areas are a consequence of the negative associations which component 3 has with social class and also reflect the relatively high number of children aged 0–14 found in many of the newer middle-class private housing estates. That this is so can be seen in the contrast between different areas within the private estate in the south of the town. St Michael's e.d.s 25 and 26, an area of very high social class and very few children, has markedly higher scores than St Michael's e.d.s 21 to 23 and 27 and 28, an area of similar social class but with many children.

This need to recognize more than one underlying factor in the explanation of the spatial distribution of the scores of component 3 is a reflection of the smaller amount of variability which this component accounts for and the greater degree of ambiguity which surrounds its interpretation. Interest in this, and the fourth component, therefore centres primarily on those areas with the most extreme scores in the direction indicated by the most highly associated variables. In the case of component 3, these are the negative associations with large houses and with shared households. The actual scores for individual enumeration districts show that the most extreme negative scores, isolating such subdivided areas, are found in the rooming-house areas and in the low-class subdivided areas. Within Bishopwearmouth, for example, Bishopwearmouth e.d. 29 has a score of -3.69, St Michael's e.d. 24 has a score of -2.35, e.d. 19 scores -2.06; in Hendon, Park e.d. 4 scores -2.22, e.d. 16 scores -3.40; in Roker e.d. 28 scores -1.21; in Monkwearmouth, Monkwearmouth e.d. 16 scores -1.65 and e.d. 21 scores -1.36. These include by far the most extreme scores in the whole range of values and since they cover rooming-house areas on the one hand and low-class subdivided areas on the other, they illustrate that these are the two types of area which are consistently and most strongly isolated by the third component.[1]

Component 4 scores

The pattern of component 4 shown in Fig. 4.10 is a complex one: high values are found in the rooming-house areas and in certain of the peripheral council housing estates; low values are found in the high-status areas of St Michael's and Fulwell and in three of the council estates (Farringdon, Hylton Castle and Hylton Red House).

To understand this pattern a number of elements related to age structure have to be invoked rather than a single dominant factor as was possible with

[1] Compare the results of the simple ecological analysis, which likewise isolated these four areas of Bishopwearmouth, Hendon, Roker and Monkwearmouth as being rooming-house or subdivided areas. See above, pp. 122–3.

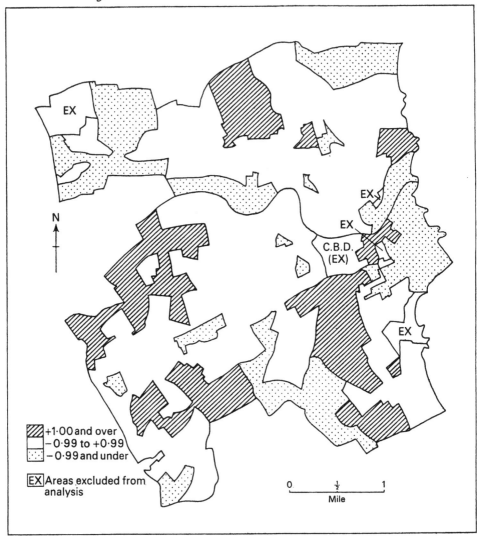

Fig. 4.10. Distribution of component 4 scores

components 1 and 2. The first element isolates the age structure of the rooming-house areas. This rooming-house syndrome is seen in the positive associations with rented furnished accommodation, large houses, females at work, young females married and young adults. All these associations might be expected to produce high scores in rooming-house areas and indeed both the Bishopwearmouth and the Roker areas stand out as having very high scores of well above 1·00. In contrast to this, the non-rooming house, sub-divided, areas of Hendon and Monkwearmouth have low scores and are clearly distinguished from the rooming-house areas by component 4. Thus, unlike component 3 which scored both types alike as areas of highly sub-divided housing, component 4 draws a distinction between them, discriminating between the different age structures of the two types of subdivided housing.

The second element of component 4 relates to the age structure of the council areas. It is negatively related to the fertility ratio and positively related to women in the labour force. As was argued above, these two variables are interconnected in that women are more easily able to work when their families are at a particular stage in their life cycle.[1] It is this combination which the fourth component picks up as its second syndrome. The effects can be seen within the council house areas where diametrically different scores are found in different estates depending on the detailed age structure of the areas involved. The crucial element in determining such wide differences is the proportion of very young children within the given areas. Those areas with high proportions of children aged 0–4 have low scores on component 4, whereas those with low proportions, even though there may be high proportions of slightly older children, have high scores. It has already been argued that such detailed differences in age structure largely depend upon the length of time that estates have been built. The estates which have low scores are in Farringdon (Thorney Close e.d.s 76 and 78 to 81), Hylton Castle (Hylton Castle e.d.s 20 to 27) and Hylton Red House (Hylton Castle e.d.s 17 to 19). The remaining council estates, by comparison, have high scores, reflecting the large proportions of women at work and the relative small proportion of very young children.

The fourth component therefore draws some most discriminating patterns within the town depending on the age characteristics associated with it. The component scores isolate the rooming-house areas from both the middle-class private areas and the low-class subdivided areas, and at the same time draw dramatic contrasts between different types of council house areas in terms of their age structures.

All told, it can therefore be seen that the four components effectively differentiate in a variety of ways between a large number of complex syn-dromes within the town's social structure. They provide a means whereby council housing areas as well as private areas can be distinguished one from

---

[1] It is of some significance that of the two fertility ratios, it is 'fertility ratio 1' (using children aged 0–4) rather than 'fertility ratio 2' (children aged 0–9) which is most highly correlated (negatively) with component 4. In fact, it is the most highly correlated of all the variables. Areas in which children tend to be rather older, and thus beyond the age at which they need closest attention, will tend to have low scores on the first fertility ratio. Conversely areas with younger children will have high scores on fertility ratio 1 and few women working. This aspect of the composition of component 4 thus fits an interpretation based on poverty and age structure rather than on *urbanization*. Cf. above, pp. 165–6.

another by a single technique. The rooming-house areas are well differentiated from the subdivided twilight areas; they in turn are well differentiated from the more stable working-class areas. In the middle-class areas, differentials are drawn with respect to differences in age and fertility ratios as well as social class. In the council housing areas, a strong and consistent differentiation is made not only between the inner redevelopment areas and the peripheral estates, but also between older and newer estates within the peripheral ring. These differences emerge not only in terms of social class and varying standards of housing but also, more interestingly, in terms of persistent variations in age structure. The component scores, summarizing nearly three-quarters of the total variability, would therefore seem to be diagnostic measures of a wide range of composite factors within the town.

AREAL GROUPINGS OF FOUR COMPONENT SCORES

As the various distribution maps of the first four components show, there are distinct areal patterns to each of the components. The final objective is therefore to attempt to combine the evidence of the four sets of component scores to produce a single framework of social areas based on the evidence of all four components. The difficulties involved in such multicomponent mapping have been touched upon in an earlier section,[1] and in the present case a variety of methods of combining the four scores was experimented with, including the calculation of generalized distance measures to produce a single estimate of closeness in terms of the four scores for each enumeration district. It was found that the best method of isolating the contiguous regional groupings was the simple cross-classification of all four of the component scores. Obviously with this method, the cutting points which are selected for each set of component scores will affect the exact areas which are delineated as homogeneous, and bearing in mind the rather different frequency distribution of the four sets of scores, two different methods are used for components 1 and 2 on the one hand, and 3 and 4 on the other. For the first two components the cutting points were selected to meet the following three requirements:

(i) the cutting points should select meaningful divisions within the frequency histograms.

(ii) a sufficient number of cutting points should be made to produce a reasonable number of divisions within each set of scores.

(iii) the number of 'homogeneous' regions should be reduced to as small a number as possible consistent with (i) and (ii) above.

[1] See above, pp. 67–72.

Since the use of some form of quantiles suffers from the fact that it spreads out the central values and so suggests heterogeneity between areas where there may be little absolute difference of values, the cutting points selected for components 1 and 2 were as follows:

Component 1		Component 2	
2·00 and over		1·50 and over	
1·00 to 1·99		0·50 to 1·49	
0·00 to 0·99		−0·39 to 0·49	
−0·00 to −0·99		−0·40 and lower	

In each case these cutting points correspond to distinct breaks in the frequency distributions and, in both cases, the range between any two cutting points equals one standard deviation (Fig. 4.6).

The cross-classification of these two sets of scores produces a total of 14 different types of combinations of components 1 and 2. In other words, of the possible total of 16 types, two are not found. These two are both combinations of very high scores both on component 1 and component 2. Such a combination would be most likely in rooming-house areas and the fact that Sunderland does not have as extensive a rooming-house area as might be found in larger towns can be held to account for the absence of this combination.

This cross-classification of the first two components accounts for nearly 60 per cent of the total variability in the universe of data and therefore, in itself, provides a discriminating breakdown of the town into its component parts. The rooming-house areas, for example, are well defined. They fall within the categories 2*c* and 2*d*, in Fig. 4.11, in which high social class is combined with poor housing. These categories form two well-delineated areas in the north and south of the town. Areas 46, 62, 64 and 65 comprise the heart of the southern rooming-house area with areas 63 and 61, in category 1*b*, forming marginal zones. In the north of the town area 18, in category 2*c*, comprises the rooming-house area, with area 19, in category 2*b*, being a marginal zone. In both cases, these designations accord well with knowledge of the areas. Likewise other areas produce valid sub-areas. Categories 1*a* and 2*a* designate the higher class areas which are found in the southern wedge of the town and, in the north, especially close to the sea coast. The stable working-class areas fall within category 3*c*, which is found especially in Pallion. The subdivided working-class areas, in contrast, are in categories 4*c* and 4*d*, and are found principally in Hendon and Monkwearmouth.

Inevitably, because only a few cutting points have been selected, a few areas fall within anomalous categories. Area 72, for example, is included within the rooming-house category (2*c*) and is the only such area falling outside the recognized rooming-house districts. Examination of the actual component scores, however, shows that this is an accident of the particular

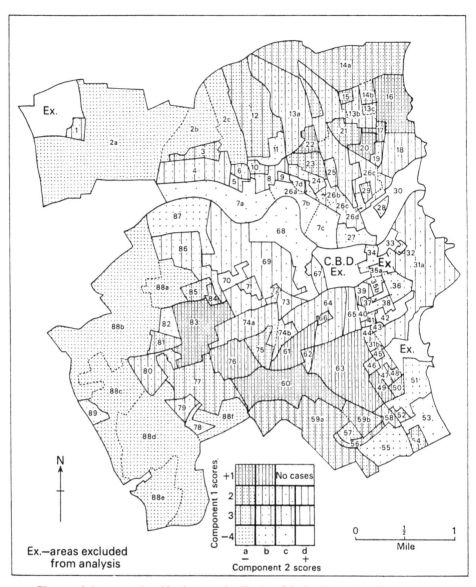

Fig. 4.11. Sub-areas produced by the cross-classification of the first four principal components.

(i) Due to its small size, the map shows only the scores of components 1 and 2.

(ii) Component 1 scores are differentiated by the spacing of the vertical lines (closely spaced lines representing high scores): component 2 scores are differentiated by the density of dots (closely spaced dots representing low scores). The cutting points used are the same as those enumerated in Figs. 4.7 and 4.8. The top left-hand box of the key (category 1a) can therefore be taken as roughly signifying high social class and good housing; the bottom right-hand box (4d) as low social class and poor housing.

(iii) Areas outlined by solid lines are those produced by the cross-classification of components 1 and 2. Dotted lines show the subdivisions made necessary by the additional consideration of the scores of components 3 and 4.

(iv) The numbers shown in each of the areas produced by this cross-classification are reference numbers and the scores of all four components for each of these numbered areas are given in Appendix E.

cutting points used. The area is comprised of only a single enumeration district which has scores on components 1 and 2 of 1·02 and 0·74 respectively. The true rooming-house areas, by contrast, all have much higher scores on both components, well above 1·00 in both cases. One typical rooming-house enumeration district within area 64, for example, has scores of 1·68 and 1·15. Area 72 thus forms an anomaly since it just barely falls within the cutting point selected for component 1 and is more rationally linked with the adjacent areas of 69 and 71 rather than with the rooming-house areas. This anomaly, and it is the most blatant, is a consequence of the blurring of differences which is inevitably produced by introducing rigid cutting points.

Indeed this and other less obvious anomalies are removed when account is taken of the scores of components 3 and 4. By adding these scores to the composite map the age structure, household composition and housing sub-division parameters, with which the two components are associated, add further differentials to the sub-areas produced by the first two components. Since components 3 and 4 account for a much smaller percentage of the total variation only their extreme scores need to be taken into account in the further building up of the composite map. In consequence the scores for these two components were expressed in various trial combinations of three divisions—an extreme positive, an average and an extreme negative. These trial divisions were based on standard deviations, sextiles, quartiles and tritiles. By superimposing these values on the 'homogeneous' regions of the first two components, it was possible to isolate those regions in which either component 3 or 4 showed both high and low scores within the same area. Such areas were considered to be heterogeneous and evidently needed to be subdivided further. The use of tritiles produced thirteen such heterogeneous areas (which included all those previously produced by the other methods) and these thirteen areas were therefore broken down into sub-areas so as to make them homogeneous in terms of the evidence of components 3 and 4 as well as the first two components. These further subdivisions are shown in Fig. 4.11 by dotted lines. The final total of areas produced in this way by the superimposition of all four components is listed in Appendix E. The areas are grouped into major categories in terms of their scores on the first two components while their scores for components 3 and 4 are noted separately. It can be seen that within the broad classification of components 1 and 2 there is a general consistency of the scores of components 3 and 4. In category 1 *a*, for example, components 3 and 4 tend to score low and high respectively, and so forth. The exceptions to this general pattern in all cases add to an understanding of the areas concerned and these exceptions are briefly discussed in Appendix E.

To summarize the main features of this composite areal grouping of the four sets of component scores, the principal characteristics of the various areas which have been delineated can be listed (see also Fig. 4.11):

(*a*) Category 1*a*
These are highly rated areas of good housing including the residential areas of St Michael's and the seaboard area of Roker, in addition to newer housing areas in the west of the town.

(*b*) Category 1*b*
This largely comprises areas of older housing of the nineteenth century which has not as yet been greatly subdivided. To this extent it is marginally rooming-house in its characteristics as measured by components 3 and 4. It also includes, however, two areas of 1930s private housing (areas 22 and 76) which have high class scores, but smaller and less well-equipped houses than those in category 1*a*.

(*c*) Category 2*a*
This is a lesser version of 1*a*. It comprises newly built private estates which tend to have higher proportions of children than 1*a*.

(*d*) Category 2*b*
These are areas of similar social class to 2*a*, but with houses of earlier date and a much older age structure with very few children.

(*e*) Category 2*c*
This is the classic type of rooming-house area with large nineteenth-century houses now subdivided and occupied by a socially mixed population, together with unsubdivided houses occupied by people of higher social class. Scores on components 3 and 4 are consistently very low and very high respectively, pointing to the age structure and social characteristics of the typical rooming-house area.

(*f*) Category 2*d*
This is a version of 2*c*.

(*g*) Category 3*a*
This includes parts of the inner ring of council estates. Area 77 is a mixture of private and council housing and its scores are therefore largely meaningless.

(*h*) Category 3*b*
This includes both the more stable working-class areas of slightly higher social class, usually housed in small cottages which have been redecorated

and improved (for example area 47) and older council estate areas such as Ford (area 86) and Newcastle Road (area 24). Again area 25 is a composite area and can be ignored.

(*i*) Category 3*c*
This is a stable working-class area with little, if any, subdivision of houses, most of which are in the form of small single-storey cottages.

(*j*) Category 3*d*
These are highly subdivided areas somewhat akin to the rooming-house areas, but with a much more homogeneous and low-class population, often of young married couples and poor families.

(*k*) Category 4*a*
This comprises the latest wave of council estates on the peripheral areas of the town. Within this category, components 3 and 4 draw a distinction between those areas with a very young age structure such as Farringdon (area 88*e*), and those in which families are generally older such as Thorney Close (area 88*d*).

(*l*) Category 4*b*
These are areas of local authority housing built in the central parts of the town such as the newly built flats in Monkwearmouth (area 27), and in Hendon (area 36), or the 1930s flats in the East End (areas 34 and 36*a*). It also includes parts of the older council estates on the periphery of the town.

(*m*) Category 4*c*
These are very dilapidated private housing areas close to the industrial areas of the river with poor housing interspersed with industrial sites. Also included is an area in the south of the town (area 53) which is located at the site of nineteenth-century industrial developments in the area of Ryhope, and which has subsequently been incorporated into the town.

(*n*) Category 4*d*
This is composed of highly subdivided housing in Monkwearmouth and Hendon much of which once comprised residential areas of higher social standing and has subsequently become the twilight zone which is scheduled for redevelopment, and parts of which have been demolished since the 1961 census.

These are the principal subdivisions of the town which are produced by the cross-classification of the first two components. The evidence of components 3 and 4, which is incorporated in Appendix E, adds additional differentiation based on a variety of demographic and social aspects. In all, they provide the basis for a framework of regions which are internally homogeneous with regard to a great range of social and economic data.

SUMMARY OF COMPONENT ANALYSIS

The use of component analysis has therefore carried the analysis of Sunderland forward some considerable way. By studying the combinations of variables which composed each of the principal components a good deal of additional insight has been gained into the types of associations which exist between individual variables. In this way, for example, it was possible to tease out the relationships between such factors as fertility ratios and the proportion of women in the labour force. Also, by studying the respective contribution of individual variables it was possible, as a by-product of the component analysis, to isolate those variables which were most diagnostic in accounting for the overall sets of associations within the total universe of the data. It has already been suggested in chapter 2 that there are promising parallels between the few urban studies which have used the technique to date.[1]

Secondly, in translating these components into areal scores, and finally combining the sets of scores into composite sub-areas, meaningful subdivisions of the town have been produced. In view of the earlier discussion in chapter 3, an especially useful feature of this process has been the technique's ability to draw valid distinctions between the various areas of local authority housing in which differentials have to be drawn in terms of age characteristics rather than the classic indexes of social class or housing characteristics. The analysis of sub-areas has therefore provided a sampling framework within which further research can be developed. In the two succeeding chapters this is done by using this framework to isolate meaningful areas within which to examine the effects of varying social and physical environments on the development of attitudes towards education.

[1] In addition to the British studies noted above, there is a growing number of multivariate studies of towns elsewhere. The published works include: C. F. Schmid and K. Tagashira, 'Ecological and demographic indices: a methodological analysis', *Demography*, 1 (1964), 194–211; B. J. L. Berry and R. A. Murdie, *Socio-economic correlates of housing condition*, Metropolitan Toronto Planning Board, Urban Renewal Study (1965); F. L. Jones, 'A social profile of Canberra, 1961', *Australian and New Zealand Journal of Sociology*, 1 (1965), 107–20; F. L. Sweetser, 'Factorial ecology, Helsinki, 1960', *Demography*, II (1965), 372–85; C. G. Janson, 'The spatial structure of Newark, New Jersey', *Acta Sociologica*, 11 (1968), 144–69. Only a single truly cross-cultural comparison has been attempted: F. L. Sweetser, 'Factor structure as ecological structure in Helsinki and Boston', *Acta Sociologica*, 8 (1965), 205–25.

PART THREE

APPLICATION

CHAPTER 5

ATTITUDES TO EDUCATION: A TEST OF THE MULTIVARIATE SUB-AREAS

If the delimitation of social areas is to become more than statistical gymnastics it has to be developed as a step to further research. The multivariate analysis of Sunderland has already thrown some light on the internal structure of the town, but one of its main objectives was to provide a sampling framework of meaningful areas which could be used to examine some aspects of the interplay between social characteristics and the urban milieu. In this chapter this is done by using the pattern of component scores to isolate seven areas within the town in which questionnaire fieldwork is used to examine the development of attitudes to education. Such attitudes can be considered as an index of social processes operating within an urban environment, and their study therefore serves both as a test of the areal sub-units which have been produced by the component analysis and as a piece of related further research in its own right. In this chapter the results of the fieldwork are presented while the more general aspects of these results are discussed in the following chapter.

The patterns of the scores of the first four components have demonstrated that there are certain regularities in the distribution of social phenomena within the town. The sub-areas which have been delimited are homogeneous only with reference to the particular sets of cutting points which have been used in the analysis so far. Only in a few cases are the lines between adjacent areas hard and fast boundaries. There are certain divisions within the town which demarcate widely different types of area. The north–south boundary between the highly rated sector of St Michael's and the low–class areas towards the sea coast is one such division. More usually, however, the boundaries are less distinct. Therefore, in place of the detailed breakdown of the town shown in Fig. 4.11, a more generalized pattern of sub-areas can be devised by examination of the component score combinations. This generalized pattern is shown in Fig. 5.1. Here the objective has been simply to combine areas which show no marked dissimilarities between the four sets of component scores. The discrete areas of high and low component scores have been used as nodes between which boundaries have been drawn.

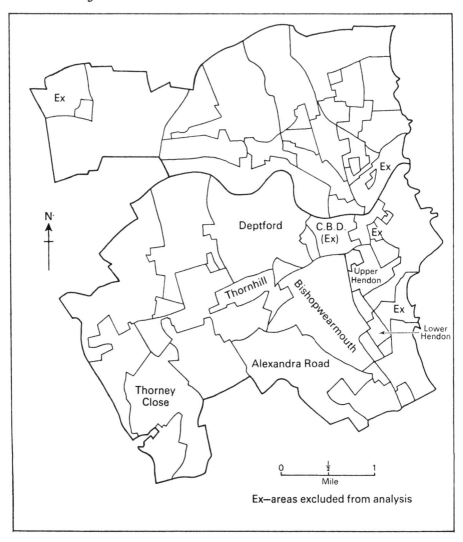

Fig. 5.1. Generalized sub-areas isolated by component analysis
The seven study areas are named

THE SURVEY AREAS

Within these broad groupings of sub-areas, seven areas have been selected covering a variety of those types of urban areas which are suggested by the component analysis. For convenience the following names and approximate designations may be given to these areas:[1]

(*a*) Deptford: stable working-class area
(*b*) Upper Hendon: highly subdivided working-class area in the 'twilight' zone
(*c*) Lower Hendon: stable skilled working-class area
(*d*) Thorney Close: council-house area with predominantly older children
(*e*) Bishopwearmouth: rooming-house area
(*f*) Thornhill: middle-class area
(*g*) Alexandra Road: upper-class area.

The descriptions are merely convenient shorthand designations which make no attempt at precision. The term 'upper class' for example, simply serves to indicate that Alexandra Road is an area of higher social class than Thornhill which is called 'middle class'. The designations thus attempt to provide easily remembered characteristics for each area and to place them in the context of the analysis of the social areas of the town. To give a more precise picture of each of the seven areas they will be examined in terms of the variables used in the previous chapter and of the component scores. For each area the history, housing conditions, social class and age structure will be examined briefly.

Deptford is an older densely settled working-class district which was built in the nineteenth century. Parts of the area date from before 1850, but most of the houses were built in the period between 1860 and 1880 and almost all of them are single-storeyed terraced cottages, a form characteristic of much of the industrial housing of the town. It is the classic type of working-class district. Close to the industrial areas fringing the River Wear, it is afflicted at once by the sound of shunting railway trucks serving the coal staiths, by the smell of gasworks, and the dirt of a variety of industries, from a power station to a cement works. Its homogeneity is heightened by physical isolation since it is shut off from other residential areas by the riverside industry on three sides, and by a railway line on the fourth. The houses themselves are small and suffer from the wear and decay of their long standing. In the north

---

[1] The actual composition of each area is as follows: Deptford—Bridge Ward, e.d.s 32–40; Upper Hendon—Park Ward, e.d.s 6–16; Lower Hendon—Hendon Ward, e.d.s 8–10; Thorney Close—Thorney Close Ward, e.d.s 68–73; Bishopwearmouth—Bishopwearmouth Ward, e.d.s 29–34, 37–8, and St Michael's Ward, e.d.s 20, 24; Thornhill—Thornhill Ward, e.d.s 42–52; Alexandra Road—St Michael's Ward, e.d.s 21–2, 25–7.

the area is scheduled for demolition, and some houses have already been pulled down. Many of the houses in the south, however, have been renovated and modernized within the limits of the ability and budget of the do-it-yourself householder, but even the facade of modernity cannot hide or compensate for the rather cramped living quarters. The energy spent on modernization is partly a reflection of the fact that increasing numbers of the occupiers are buying the houses so that the market for rented property within the area has been severely restricted in recent years. Indeed the 1961 census data for the relevant enumeration districts show that 40·1 per cent of all households were owner-occupied.

Upper Hendon is of a similar age, but is composed of very different types of houses and has a markedly different history. The changes in the residential status of the area were examined in chapter 3 when it was seen that the area was originally built as a good residential district and has subsequently suffered invasion, progressive decay and falling status as succeeding populations have encroached on the area following the development of industry southwards along the sea coast.[1] The houses are two- and three-storeyed terraces typical of much of the Middle and High Victorian town house of middle-class areas. Architecturally they make quite an impressive series of terraces, but the progressive subdivision and the overcrowding that this has brought with it have caused neglect, decay and gross disrepair. Virtually every house is tenemented and lived in by two or more household units. Only exceptionally are individual houses single dwelling units and, occasionally, such houses have preserved something of the original character of the area.

Lower Hendon is in many respects similar to the stable working-class area of Deptford. While it was mostly built at a later date (the late 1890s), the type of housing is almost identical, consisting of rows of single-storeyed cottages, but rather larger and more substantial than those in Deptford. In the southern part of the area streets of two-storeyed dormer-window houses alternate with streets of single-storeyed cottages. While the rating values of the two areas are comparable, those in Lower Hendon tend to be marginally higher throughout, reflecting the better physical conditions of the houses and the fact that most have been more extensively modernized and more carefully looked after. Likewise, the land use of the area is slightly different. While there is adjacent industrial development it is of a less noxious character and more discreetly placed. As in Deptford, the corner shop is a ubiquitous feature of the area. Lower Hendon is therefore a rather better version of Deptford, but very different from the subdivided terrace housing of Upper Hendon. Of the houses in the area some 66 per cent are owner occupied.

[1] 'Upper Hendon' comprises most of the area studied in chapter 3 as 'Hendon'—the subdivided lower-class area. See above, pp. 126–7.

Thorney Close is a post-war council housing estate built in the mid 1950s. Architecturally, it is a collection of low-density, sub-terrace houses of the type found widely in local authority areas of its period. Built largely to accommodate people from clearance areas within the inner parts of the town, it houses a population of predominantly low social class and, having been occupied for a number of years, there is a relatively low proportion of children aged under 5, but high proportions in the quinquennial groups between 5 and 20.

Bishopwearmouth comprises the area of the southern rooming-house district about which much has already been written (see above, pp. 123–6). At one time the inner apex of the high status area of the town, the area has undergone profound changes in occupance since it was first built. With the exception of some of the imposing Georgian terraces of the central part of Sunderland, the houses within Bishopwearmouth are the largest and most impressive in the town. Most of its streets are composed of tall three- or four-storeyed terraces and there are also some fine squares and large detached houses standing in extensive grounds. As has been shown earlier, the difference between this area and the twilight zone of Upper Hendon is that in Bishopwearmouth fewer of the houses have been subdivided and those which have are usually occupied by a population of higher social status and with much smaller, if any, families. Of the dwellings, 35 per cent are owner occupied, while 14 per cent are rented furnished. Indeed, while the buildings themselves are of fairly uniform type, the social composition of the population is very mixed. Apart from the, generally, high status families who own and live in undivided houses, there is a large body of students and of young newly married couples drawn from a range of social classes, but predominantly of higher social status than the population of Upper Hendon.

Thornhill is a much more uniformly residential area than most of those already considered. There is a marked progression from smaller housing in the eastern part to larger and more substantial housing in the west. Some of its streets alternate between rows of cottages, many of which are lived in by elderly people, and substantial two-storeyed terrace houses lived in by younger families with children. Most of the area was built around the turn of the century, but parts date from the mid-war period. The solidly residential aspect of the area derives partly from the more substantial houses and partly from the absence of corner shops and local industry which differentiates this area from the working-class areas of Deptford and Lower Hendon. Unlike these areas, within Thornhill there is a clear separation between residential and non-residential land uses and this is reflected in its much more middle-class population. Within the area 75 per cent of houses are owner occupied.

TABLE 5.1. *Social and physical characteristics of the seven study areas*

(*a*) Average component scores

	Component			
Area	1	2	3	4
a Deptford	0·317	0·939	1·358	−0·385
b Upper Hendon	−0·069	1·616	−0·558	−0·184
c Lower Hendon	0·847	−0·024	1·256	−0·061
d Thorney Close	−0·359	−0·881	−0·005	0·809
e Bishopwearmouth	1·510	1·204	−1·470	1·767
f Thornhill	1·536	−0·139	0·549	−0·170
g Alexandra Road	2·451	−1·002	−0·902	−1·102

(*b*) Housing conditions

Area	Persons per room	Percentage households at > 1½ per room	Percentage households with all four amenities	Percentage households sharing dwellings	Percentage building type 1	Percentage owner occupied
a Deptford	0·86	8·6	23·2	8·7	80·5	40·4
b Upper Hendon	0·95	13·9	14·0	45·0	64·4	22·6
c Lower Hendon	0·71	1·4	64·2	2·0	91·4	66·0
d Thorney Close	1·02	7·5	99·7	0·3	85·5	0
e Bishopwearmouth	0·62	5·2	51·1	29·9	57·0	35·2
f Thornhill	0·62	1·1	78·3	7·9	86·4	74·8
g Alexandra Road	0·58	0·2	99·3	0·1	99·0	94·5

(*c*) Social class characteristics

Area	Percentage Class I	Percentage Class II	Percentage Class III	Percentage Class IV	'Class score'
a Deptford	3·8	8·2	53·5	34·3	181·5
b Upper Hendon	3·2	7·6	50·2	38·9	175·0
c Lower Hendon	7·4	16·4	46·3	29·9	202·0
d Thorney Close	1·9	7·1	47·9	43·0	168·0
e Bishopwearmouth	21·3	25·1	33·4	20·2	247·6
f Thornhill	16·9	27·1	39·5	16·5	244·2
g Alexandra Road	51·1	30·7	16·0	2·1	330·4

NOTES. *i* Areas are defined in text, p. 189.
ii Social class definitions are provided in Table 4.1.
iii Housing definitions are given in Table 4.1, except for 'Building type 1' which comprises wholly residential permanent buildings containing only one dwelling unit.
SOURCE. Census, and component analysis.

Alexandra Road, the final area, is the present-day nucleus of the high status sector of the town, which has moved progressively outwards to the southern boundary. The north-western parts of the area were built between 1935 and 1939, while the south-east is of post-war date. The older areas consist of detached and semi-detached housing built at low density in pleasant residential surroundings. The adjacent post-war areas tend to be of smaller houses most of which are semi-detached and which tend to house a greater number of children and young families. This difference is reflected in the rather different scores between the north-western and south-eastern areas on components 3 and 4. Of the dwellings in the area, 95 per cent are owner occupied.

To examine the social, housing and age characteristics of each area in more detail a selection of variables has been drawn from the census enumeration district data and average scores are shown in Table 5.1. These average scores provide a synoptic picture of the socio-economic characteristics of the seven areas. Component 1, which is most closely related to social status, shows that Thorney Close has by far the lowest score, followed by Upper Hendon. Next in rank are, respectively, the two working-class areas of Deptford and Lower Hendon. Bishopwearmouth and Thornhill are each very similar in score, while Alexandra Road stands out as by far the most highly scored area. In component 2, which is negatively associated with good housing, the two subdivided areas of Upper Hendon and Bishopwearmouth understandably have the highest scores. Deptford has slightly less high scores and is followed by Lower Hendon and then Thornhill. Thorney Close and Alexandra Road—the council estate and the high-class private estate— have the lowest scores, reflecting their good housing facilities. Component 3, which is negatively associated with subdivision, isolates Bishopwearmouth and Upper Hendon. High positive scores are found in the two working-class areas of Deptford and Lower Hendon. Component 4, negatively associated with a young age structure, differentiates between Bishopwearmouth, which has high scores, and Upper Hendon, Alexandra Road, and Deptford, which have low scores.

The average values of individual variables associated with housing characteristics in each area provide more detailed information which supports these component score differentials drawn between the areas. The most well-provided areas in terms of amenities are clearly Thorney Close and Alexandra Road, while the subdivided area of Upper Hendon is poorly housed, with only 14·0 per cent of dwellings having all four amenities. Deptford, too, is seen to be a badly housed area. The figures of density of occupation show that while Thorney Close has a high average density, only 7·5 per cent of households are living at over 1½ persons per room. By comparison Upper Hendon has a lower average density, but a much higher proportion of overcrowding. This is a reflection of the admixture of crowded subdivided dwellings and less congested single-family dwellings in the area. Similarly, the figures for the rooming-house area show the same signs of diversity. While Bishopwearmouth has an average density of only 0·62 persons per room, 5·2 per cent of the households live at over 1½ persons per room. This compares with only 1·1 per cent in Thornhill which has exactly the same average density. This is a consequence of the very different type of social and housing composition of the two areas. The degree of subdivision in Upper Hendon and Bishopwearmouth is clearly brought out. Upper Hendon has 45 per cent of its households in shared accommodation and Bishopwearmouth has almost 30 per cent. Likewise, whereas the other five areas have 80 per cent and more of their houses in the census category 'building type 1', Bishopwearmouth has only 57 per cent and Upper Hendon has 64·4 per cent.

The social class characteristics confirm the scores of the first component, and add to the brief description of the seven areas above. The average 'class score' ranks Thorney Close as the lowest area, followed by Upper Hendon, Deptford, Lower Hendon, Thornhill, Bishopwearmouth and, well above the others, Alexandra Road. The breakdown of these average scores, however, shows that the social composition of the areas is not a simple progression from low-class area to highly rated area. Thorney Close has almost as large a proportion of people in class IV as in class III with very few in the higher classes.[1] Deptford and Upper Hendon have very similar proportions in all four class groups with slightly over a half of the total in class III. Lower Hendon is only marginally different with a general upward shift in the social composition: more in the two upper groups and correspondingly fewer in the lower two. Thornhill represents a further, and more marked, step on this upward progression with 44 per cent in the two upper groups. Alexandra Road forms the opposite extreme to the council estate of Thorney Close, since it has over half of the total in class I and over 80 per cent in the first two social groups. The most interesting distribution, however, is in Bishopwearmouth where, as was suggested above, the social composition is indeed seen to be very mixed. Unlike the other six areas, no one social group stands out prominently. One-third of the total is in class III, but the remaining two-thirds are divided almost equally between classes I, II and IV. This compares very differently with the distribution in the other subdivided area in Upper Hendon where classes I and II are barely represented at all.

The housing conditions and social composition of the areas are thus fairly clearly defined and in themselves add validity to the component analysis

---

[1] For a definition of these classes, based on socio-economic groups, see the list of variables in Table 4·1, p. 136.

which isolated these seven areas. The final characteristics which will be examined before considering the questionnaire survey are the age structure and the fertility ratios of the areas, since these features play a crucial part in the argument of the next chapter concerning the development of attitudes to education.

TABLE 5.2. *Demographic characteristics of the seven study areas*

(a) *Age structure* (percentage of population in each quinquennial age group)

	Area						
Age group	a	b	c	d	e	f	g
0– 4	9·49	14.80	8·80	8·15	8·79	6·87	8·22
5– 9	7·09	7·76	6·50	10·00	4·69	6·23	7·55
10–14	5·98	5·87	6·63	16·50	5·10	5·84	7·08
15–19	5·23	4·72	5·33	10·52	6·26	5·92	5·33
20–24	6·89	9·70	5·32	5·90	10·60	5·88	3·87
25–29	8·37	10·64	7·52	4·39	9·07	6·37	5·75
30–34	6·80	7·59	6·68	4·84	6·08	5·63	8·29
35–39	6·16	6·11	6·70	7·56	5·42	5·87	7·59
40–44	5·48	4·12	6·39	8·29	4·94	6·30	7·59
45–49	5·85	4·81	5·83	7·15	5·54	7·44	7·59
50–54	5·87	4·81	6·80	6·11	6·16	7·07	7·55
55–59	6·53	4·69	6·99	3·61	5·24	6·72	7·40
60–64	6·12	4·38	6·79	2·55	5·98	6·42	4·41
65–69	5·04	3·46	5·34	1·69	5·05	5·60	4·38
70–74	4·28	3·05	3·54	1·16	4·77	5·21	3·40
75–79	2·60	2·07	2·89	0·81	3·25	3·37	2·38
80–84	1·41	1·04	1·36	0·41	2·16	2·21	1·02
85 and over	0·78	0·35	0·56	0·34	0·90	1·05	0·69

(b) *Fertility ratios* (children aged 0–9 as percentage of females aged 15–44)

a	Deptford	86·98
b	Upper Hendon	104·43
c	Lower Hendon	79·03
d	Thorney Close	85·01
e	Bishopwearmouth	61·35
f	Thornhill	71·38
g	Alexandra Road	76·93

NOTE. Areas listed in part (a) are as given in part (b).
SOURCE. Census.

The proportion of persons in quinquennial age groups in each of the areas is shown in Table 5.2. These average values conceal a certain degree of variation in age structure between the different enumeration districts in every area apart from Thorney Close where there is a great similarity of age structure from one enumeration district to another. The average values, however, do provide an overall picture of each area. It can be seen that Thorney Close has two pronounced concentrations, one at 10–15, the other from 35 to 50. There is a large number of children, of all ages, in addition to the bulge in the young teenage group. There are very few elderly people. Upper Hendon, likewise, has a large number of young children. The bulge in this case occurs in the age group 0 to 4. There is a second bulge between the ages 20 to 35, which emphasizes the fact that this is an area of essentially young, married persons with large families. Unlike the council estate, there is a relatively large number of elderly people in the population. Deptford and Lower Hendon each have remarkably similar overall age distributions. Large proportions of persons of middle and old age are combined with a deficiency between the ages of 15 to 30. The two areas differ only in that Deptford has rather more young children than has Lower Hendon, and a rather greater proportion of persons aged 25 to 30. This would suggest a greater number of young families within this area. Thornhill shows a rather different pattern again, with fewer young children, but a larger proportion of young persons aged 15 to 20, and similar proportions of elderly and middle-aged persons. More of the families in Thornhill are thus in the static stage of the family cycle with grown or nearly grown children. There is also a social class difference superimposed to some extent on top of this explanation of the age structure in Thornhill, since children in the late 'teens in such a middle class area tend to stay at home longer and marry later than do children in a working-class area. Alexandra Road can be seen to be a much younger area overall. There are two broad bulges in the age distribution: one of young and youngish children, the other of middle-aged persons between the ages of 30 and 60. There are fewer elderly persons than in Thornhill and, unlike Thornhill, a marked deficiency in the ages between 15 and 25. Finally, Bishopwearmouth shows elements of the classic rooming-house district. There is a marked bulge between the ages 20 and 30 and, with the exception of children aged 0 to 4, very few children and young persons. The excess of children in the age range 0 to 4 is explained by the existence within the area of a certain number of young, recently married couples with young families. While not in accordance with the classic rooming-house area traits, this is a reflection of the housing shortage in Sunderland and the fact that this area is one of the few which offers a wide market in rented accommodation.

The fertility ratios of the areas show similar patterns to the age structure. The fertility ratio measured by the number of children aged 0–9 as a percentage of women aged 15–44, shows that Upper Hendon has by far the greatest ratio. Thorney Close and Deptford (the council estate and the stable working-class area) have the next highest ratios. Lower Hendon, with its smaller percentage of children, has a lower fertility ratio than Deptford in spite of the many similarities between the two areas. Not surprisingly, the lowest ratio is found in the rooming-house area of Bishopwearmouth, as indeed the pattern of component scores in the last chapter suggested.

These average values of the age structures and fertility ratios conceal differences within the areas. The one which most calls for comment is the considerable difference between the older and the more recent residential areas in the Alexandra Road area. The longer-established part to the east has a somewhat older age structure with lower fertility ratios than the newer parts, and this was reflected, and commented upon, in the component analysis of the last chapter.[1]

QUESTIONNAIRE SURVEY: ATTITUDES TO EDUCATION

These seven areas were therefore selected to examine the sociological effects of areal differentiation within the town. The dependent variable which was used was parental attitudes towards the education of their children. This parameter was selected because of the well-documented association between social class and educational achievement. In Sunderland this can be demonstrated in a number of ways. As one example, the results of the 11-plus examinations held in 1962 and 1963 illustrate the pattern of average intelligence scores within sub-areas of the town (Fig. 5.2). Records of all the boys taking these examinations provide evidence of the measured intelligence quotient of each candidate. In all, over three thousand individual I.Q. ratings were plotted and, while some of the smaller sub-areas of the town contained too few candidates to enable reliable average figures to be calculated, most of the residential areas could be allotted such scores. The pattern which emerges is very similar to the familiar pattern of status areas which has been isolated within the town, similar, in other words, to the distribution of social class, rating valuations and the other social class parameters. Indeed, the correlation between social class and the I.Q. values is as high as $r = 0.617$. The same high positive association exists between parents' social class and the type of school (whether grammar, technical or secondary modern) which they choose for their children; the highly rated areas having consistently large proportions wanting their children to attend grammar schools. Such associations between social class on the one hand, and educational achievement and aspiration on the other are to be expected in the light of the work that has been done on the sociology of education.

That parental encouragement affects the progress of a child throughout its formal education cannot be doubted. One of the most recent of many demonstrations is seen in Douglas' work, where it is shown that even when social class is held constant, the child whose parents show an interest in his

[1] See above, p. 175.

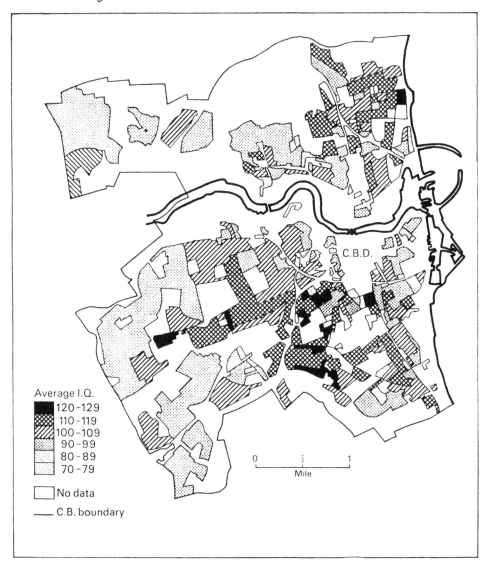

Fig. 5.2. Average intelligence quotients of children

Source: 11-plus examination results, 1962 and 1963

school work tends to achieve more in relation to his I.Q. level than the child whose parents are uninterested. As Douglas comments: 'The influence of the lower level of parents' interest on test performances is greater than that of any other of the three factors—size of family, standard of home, and academic record of the school—which are included...and it becomes increasingly

198

important as the child grows older.'[1] The development of parental attitudes is therefore of some significance in educational fields. In the field of sociology, one of the principal interests has been in the close positive association between social class and attitudes towards education. The higher the parents' social class, the more likely are they to take an active interest in their children's education and to encourage them to do well at school.[2]

Social class, however, is a theoretical construction. One fact which emerges from the work of Centers, for example, is that people do not generally think in terms of an all-embracing model into which their fellow men can be fitted. If they do think in terms of a model, then a single model is not universally recognized, but different ones are used by different individuals. As Bott argues, 'the individual constructs his own notion of social position and class from his own segregated experience of prestige and power and his imperfect knowledge of other people's...He creates his own model of the class structure and uses it as a rough-and-ready way of orientating himself in a society so complex that he cannot experience directly more than a tiny bit of it.'[3] The individual's conception of social structure and his place within it is formed with reference to a particular social environment which can be thought of as his 'social world'. This social world is, for most people, determined by face-to-face contacts at work, in the home, in the neighbourhood and in the various facets of the individual's leisure and working life. If attitudes towards education correlate with social class, it seems reasonable to assume, therefore, that this is not a result of a conscious acceptance or awareness of being in a particular social class, but rather because of certain forces and experiences to which people of a similar class each tend to be exposed.

The neighbourhood, or the immediate physical and social environment in which people live, is an important source of some of these common forces which influence the development of attitudes towards education.[4] It is not suggested that this is the only or the most important factor to consider. The relative importance of the home, the workplace, the club, the neighbourhood, the influence of relatives, and the other formative factors, in contributing to

[1] J. W. B. Douglas, *The home and the child : a study of ability and attainment in the primary school* (London, 1964).
[2] See, for example, F. M. Martin, 'Parents' preferences in secondary education', in D. V. Glass (ed.), *Social mobility in Britain* (London, 1954), pp. 160–74.
[3] E. Bott, 'The concept of class as a reference group', *Human Relations*, III (1954), 263: R. Centers, *The psychology of social classes* (Princeton, 1949), ch. 2. An interesting use of reference groups, used in a somewhat similar fashion to their use in the present study, is found in W. G. Runciman, *Relative deprivation and social justice : a study of attitudes to social inequality in twentieth-century England* (London, 1966).
[4] For recognition in the sociological literature, of the role of neighbourhood *vis-à-vis* family and personal influences, see, for example, C. M. Fleming, *The social psychology of education* (London, 1944), ch. 9.

the general process of moulding the attitudes that develop, will depend on the particular 'social world' of each individual. Obviously social class is related to many of these factors. Social class and area of residence are to a large extent coincident, as has been shown in detail above, but the pattern of one-class homogeneous neighbourhoods is by no means perfect. The effect of the neighbourhood on the development of attitudes could thus be studied by examining not only those people whose social class does not accord with the social class of the area in which they live, but by applying a geographical approach to the study of attitudes using *location* rather than social class as the principal independent variable. It might be expected that, if the area in which a person lives does in some way influence his attitudes and behaviour, then it should have some independent effect on his attitude towards education over and above those associations which are correlated with his social class considered in isolation.

To test this hypothesis, information was collected within the seven areas outlined above by a questionnaire survey conducted during July, August and September, 1963. Within each of the seven areas, the sample of population consisted of the parents of all boys who were due to sit the local authority 11-plus examination in the following year, that is in Spring, 1964. The sample universe was collected from school registers in the Sunderland Education Offices, which contain addresses of every boy due to sit the 11-plus examination (and thus include boys attending private as well as state schools and schools both within and outside the town). The 'sample' consists of all boys living within one of the seven survey areas who were due to sit the examination. Only a relatively few addresses proved to be out of date (see Table 5.3).

The sample was therefore not an attempt to provide a cross-section of the areas themselves, but aimed to achieve a certain degree of uniformity of educational experience amongst the sampled population. By selecting the parents of boys about to sit an examination which would have important consequences for their future formal education, it was intended to provide a focus for the questions asked and to eliminate at least two of the most important sources of variation which were not relevant to the research design; namely the sex and the age of the children. The different expectations which patents have as regards boys and girls is well known. By excluding the parents of girls who were about to take the 11-plus examination, this difference was eliminated. Second, the age of the child is obviously of great importance. It can have effect in a number of ways: if education is valued in terms of the financial rewards of a better job for the child, disillusionment on this score, after the child has ended formal education, may alter the parent's esteem for education; if a parent has, perhaps at considerable personal sacrifice, encouraged a child to pursue a long period of formal education only to find that subsequently the child rejects the parents' values, then parental disenchantment with the educational system is likely to result. Changes in parental attitude are therefore closely related to the educational stage which the child has reached and to whether this has resulted in favourable or unfavourable experiences in terms of the particular goals to which education was seen to lead. Thus it would have been preferable to select only those

TABLE 5.3. *Survey population in the seven study areas*

Area	Initial population	Incorrect address	Not contacted	Refusal	Interviewed	Adequate material
a Deptford	20	2	1	.	17	17
b Upper Hendon	48	8	3	.	37	35
c Lower Hendon	20	1	.	.	19	19
d Thorney Close	38	.	1	1	36	36
e Bishopwearmouth	17	1	2	.	14	14
f Thornhill	38	.	.	1	37	36
g Alexandra Road	34	.	2	.	32	31
Totals	215	12	9	2	192	188

NOTE. In each case, numbers represent one family (with a son about to sit the local authority 11-plus examination).

parents whose eldest child was about to sit the examination, so that such factors could have been controlled more fully. This however would have reduced the sampling universe to a fraction of what it otherwise was. Indeed, taking all parents with a son about to sit the 11-plus examination in the seven areas, the total sampling population was only 215. In assessing attitude scores, this total was finally reduced to 188, as is made clear in Table 5.3. This reduction of the original population was partly the result of families having moved house and partly the result of difficulty in contacting some of the parents despite repeated visits to their homes. Of the total of 194 homes visited, in only two cases was there a refusal to co-operate. In 9 cases parents could not be contacted.

QUESTIONNAIRE RESULTS

The questionnaire is reproduced in Appendix F. The questions covered a wide field and were supplemented in all cases by informal discussion of matters related to education. The topics of immediate concern are social class and attitudes towards education. The methods used to measure each of these parameters are discussed in Appendix G. It will suffice to note first that four social classes were delineated on the basis of the Registrar General's

Urban analysis

Classification of Occupation. These classes may be designated as follows: (1) upper middle class; (2) lower middle class; (3) skilled working class; (4) semi-skilled and unskilled working class.

In quantifying these groups, the first class (upper middle) was allotted a score of 1, the second class a score of 2, the third a score of 3, and the fourth a score of 4, for each individual (either male or female). Where reference is made to the 'combined' score of mother and father, this is simply an addition of the individual scores of the two parents. For example, where both parents are in class 1, the upper middle, the combined score will be 2. If one parent is in class 2, the lower middle, and the other in class 3, the skilled working, the combined score will be 5. The range of the individual scores is therefore from 1 to 4, and of the combined scores is from 2 to 8. In all cases, lower numbers represent higher classes.[1]

Second, the attitude responses were quantified by two methods as Appendix G makes clear. The first was a simple weighting of responses so that favourable replies scored 1, unfavourable replies scored 0. The second method was devised in terms of the Guttman scalogram, which has been used in quantifying a variety of types of attitude and has been successfully applied in other fields.[2] As the Appendix discussion makes clear, the scalogram technique eliminates those questions which are inconsistent with the majority of responses and ranks the consistent questions in order of their progressive discrimination as regards favourable attitudes to education.[3] The resulting set of responses form what Guttman calls a 'scalable' universe: they are, in other words, all part of a single attitude dimension. He suggests that the consistency of any scale can be measured by a 'coefficient of reproducibility' which should fall above a level of 0·90 if the scale is to be accepted. In the case of the results of the Sunderland questionnaire responses, this coefficient was 0·94. It can therefore be taken that increasing scores as measured by the pattern of scaled responses do reflect a real continuum of interest in and encouragement of their son's education by the parent respondents. Since the scale is a compound of the length of formal education which the parent wishes each boy to have, and the amount of help and encouragement which is

[1] In the subsequent presentation of the questionnaire results, tables and discussion are frequently presented in three distinct fashions which are referred to as 'father's class', 'mother's class' and 'combined class score'. These designations refer to the social grouping of the family as it is classified and re-classified using the evidence of father's occupation, mother's work before marriage and the combination of these two.

[2] S. A. Stouffer, L. Guttman, E. A. Suchman, P. Lazarsfeld, *Studies in social psychology in World War II*, vol. IV, *Measurement and Prediction* (Princeton, 1950), chs. 2–9. Two applications of the scalogram to the analysis of urban morphology and structure are found in N. E. Green, 'Scale analysis of urban structure: Birmingham, Alabama', *Am. Sociol. Rev.* XXI (1956), 8–13, and D. W. G. Timms, 'Quantitative techniques in urban social geography', in Chorley and Haggett (eds.), *Frontiers in geographical teaching*, pp. 257–61.

[3] See the scalogram reproduced in Appendix G.

offered, the scale measures the degree of aspiration that the parent shows on behalf of the boy. The two types of measure were combined in a ratio of 1:2 so as to give greater weighting to the scalogram scores. Since the maximum score on the first measure was 13, and the maximum score on the scalogram was 8, the range of final attitude scores varies between 0 and 29, with higher scores indicating greater interest in and enthusiasm for the child's education.

Turning to the results of these scores, there is a significant positive correlation between social class and educational aspiration whether one orders the families in terms of the class of the father, the class of the mother (as judged by her work before marriage), or the combined score of both mother and father. The various Pearsonian correlation coefficients are:

$$\text{Father's class} \quad r = 0\cdot523$$
$$\text{Mother's class} \quad r = 0\cdot514$$
$$\text{Combined class} \quad r = 0\cdot530$$

All three of these coefficients are significant at $P = 0\cdot01$. It is interesting to note that, while there is only a slight variation between the three coefficients, that associated with the father's social class is marginally greater than that using the mother's social class. It was expected that the correlation of attitudes with the mother's class would have been the higher of the two, especially since in most cases the interviews were with the mother herself, but also because much of the research concerned with attitudes towards education has pointed to the greater importance of the mother in determining the achievement of children in school.[1] The fact that the mother's class is based on the work of the mother before marriage may partly explain this difference in the observed results since frequently this is a less diagnostic indicator of the social class of the woman, than is the man's present occupation as index of his social class.

While significant, these correlations between social class and attitudes are perhaps somewhat lower than might have been expected. Tables 5.4 and 5.5. show that there is a considerable scatter of scores within each of the four groups of father's and mother's social class. This is particularly so in class 3 (skilled manual) in the father's class groups, and in classes 2 and 4 (lower middle and semi-skilled and unskilled) in the mother's class groups where individual attitude scores vary throughout the whole range from 0 to 29. In both classifications, however, the average attitude scores show the expected upward progression, irregular in the case of the father's classification, more regular in the case of the mother's.

In the father's classification (Table 5.4), the average scores for classes 1 and 2 are very similar. The fact that the average score for class 1 is not higher

[1] Douglas, *The home and the school*, pp. 43–4.

TABLE 5.4. *Social class and attitudes (families classified by father's occupation)*

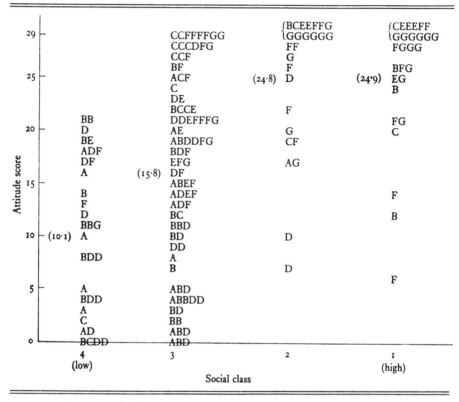

NOTES. *i* Average attitude scores for each class are shown encircled.
ii Entries represent one family from the following areas:

A	Deptford	E	Bishopwearmouth
B	Upper Hendon	F	Thornhill
C	Lower Hendon	G	Alexandra Road
D	Thorney Close		

is largely due to the fact that the ceiling of aspiration which is set by the questionnaire may well be set at too low a level to distinguish effectively between the various respondents who aspire very highly for their sons. A more rigorous set of questions might have revealed more differentiation at the top end of the scale so that, rather than having a large number of respondents all scoring the maximum 29, there would have been more in social class 1 and fewer in social class 2 scoring above this figure. Since the primary interest of the study, however, was in differentials in the middle and lower ranks of the social scale, the aim of the attitude survey was to use the upper middle or lower middle class respondents as a yardstick or a gauge

TABLE 5.5. *Social class and attitudes (families classified by mother's work before marriage)*

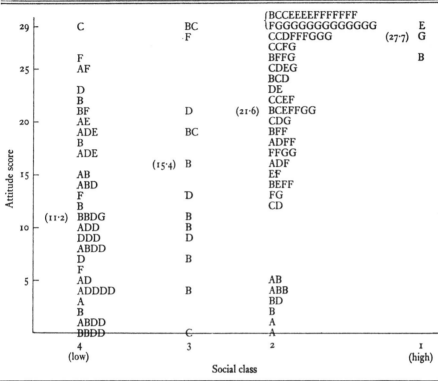

Attitude score	4 (low)	3	2	1 (high)
29	C	BC	⎰BCCEEEFFFFFFF⎱ FGGGGGGGGGGGGGG	E
		.F	CCDFFFGGG (27·7)	G
			CCFG	
	F		BFFG	B
25	AF		CDEG	
			BCD	
	D		DE	
	B		CCEF	
	BF	D	(21·6) BCEFFGG	
20	AE		CDG	
	ADE	BC	BFF	
	B		ADFF	
	ADE		FFGG	
		(15·4) B	ADF	
15	AB		EF	
	ABD		BEFF	
	F	D	FG	
	B		CD	
	(11·2) BBDG	B		
10	ADD	B		
	DDD	D		
	ABDD			
	D	B		
	F			
5	AD		AB	
	ADDDD	B	ABB	
	A		BD	
	B		B	
	ABDD		A	
	BBDD	C	A	

Social class

NOTES. *i* Average attitude scores for each class are shown encircled.
ii Entries represent one family from the following areas:

A Deptford	E Bishopwearmouth
B Upper Hendon	F Thornhill
C Lower Hendon	G Alexandra Road
D Thorney Close	

against which to measure the responses of lower-class families. The concern was not so much with the absolute totals of the attitude scores—especially since any interpretation of such totals is very difficult—but rather with the differences between respondents, particularly at the lower levels.

When the families are ranked in terms of the mother's class (Table 5.5), there is a more even gradation with increasing average scores as one moves from class 4 to class 1, but here, too, there is a large range within any one social class grouping. Both maximum and minimum scores are found within no fewer than three of the four class groups.

Many of these anomalous class scores are removed when the respondents are ranked in terms of the combined class scores of both father and mother, as shown in Table 5.6. The average attitude score is higher in each successive division of the seven class groupings. Thus at the lowest end of the social scale in which the combined class score is 8 (i.e. both father and mother are in class 4, the semi-skilled and unskilled class), the average attitude score is 7·3 out of the maximum possible 29. At combined class score 7, the average attitude score is 12·9; at class score 6, it is 15·0; and the progression continues up to class score 2 (in which both father and mother are in class 1, the upper middle class), in which the average attitude is 27·5. The interval between each average attitude score is not an even one. In the three highest groups, for example, there is little difference between the average scores. It has been suggested above why this evening-out might be expected to occur. Indeed, it is not surprising that one does not find a perfect linear relationship between the two variables, since the class and attitude scores were not devised to attempt to establish a 1:1 relationship between the two. The point to be made is that there is a very significant correlation between class and attitude as measured by the results of the questionnaire survey.

AREAL PATTERNS

The most interesting findings only emerge however when the respondents are examined in relation to the areas in which they live. The plotting of the combined class groups (Table 5.6) shows, for example, that there is at least one curious group of respondents who appear to be out of alignment, as it were, in relation to other respondents within the same social class. The most marked of these anomalies is in the class score 5 grouping in which there is a small isolated knot of respondents who have much lower attitude scores than the remainder. It is interesting to note that these respondents come largely from two of the seven survey areas. In all, there are four from Upper Hendon, the low-class subdivided area, three from Deptford, the stable working-class area, and one each from Thorney Close, the council estate, and Thornhill, the 'middle class' area. The other areas are not represented. In other words it is only in certain of the areas that people within this particular social group tend to 'underscore' in relation to their class grouping. The same sort of pattern is repeated with 'underscorers' in the other social class groupings. A very different pattern applies when the 'overscorers' are considered. Evidently, the numbers involved in any areal comparison within the respondents are small and differences of the kind noted above are not necessarily significant, but in spite of the small numbers involved consistent patterns which recur throughout the total population of the questionnaire respondents will be taken as having a certain intrinsic significance.

TABLE 5.6. *Social class and attitudes (families classified by combined occupations of mother and father)*

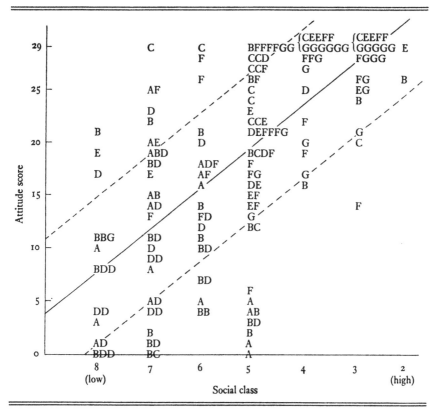

NOTES. *i* The regression line of *y* on *x* is shown, together with lines drawn at one standard error from the regression line.

ii Each entry represents one family from the following areas:

A	Deptford	E	Bishopwearmouth
B	Upper Hendon	F	Thornhill
C	Lower Hendon	G	Alexandra Road
D	Thorney Close		

As an initial index to isolate those areas which have a preponderance of 'overscorers' and those with a preponderance of 'underscorers' in relation to social class, the average class and average attitude scores of each of the seven areas may be compared, as in Table 5.7. Whereas there is a fairly regular increase in the average class score of respondents from the seven areas as one progresses from Thorney Close, to Deptford, to Upper Hendon, to Lower Hendon, to Bishopwearmouth, to Thornhill and finally to Alexandra Road, the average attitude scores for the areas clearly fall into two distinct groups with Deptford, Thorney Close and Upper Hendon falling in a low scoring group with average scores between 10 and 12, and the remaining areas falling into a higher scoring group with average scores of between 22 and 26. Certain of the areas would thus appear to underscore consistently; others would appear to overscore in relation to social class.

TABLE 5.7. *Average social class and average attitude scores*

		Average class score			Average attitude score
Area		Father's class	Mother's class	Combined class	
a	Deptford	1·71	1·71	3·42	10·4
b	Upper Hendon	2·00	1·91	3·91	11·6
c	Lower Hendon	2·22	2·71	4·93	22·7
d	Thorney Close	1·76	1·60	3·36	11·1
e	Bishopwearmouth	2·64	2·50	5·14	22·9
f	Thornhill	2·51	2·68	5·19	22·5
g	Alexandra Road	3·12	2·96	6·10	25·5

The areal patterning of this over- and under-scoring can be made clearer by returning to the graph of the combined class scores and attitude scores in Table 5.6. The regression line of y on x has been added in addition to the average scores for each of the class groups. Lines at one standard error (Sy) have been drawn on either side of this regression line. The various respondents can be divided into rough categories in relation to these lines. Those lying above the regression line can be thought of as 'overscorers' with those lying above the line at one standard error above the regression line as 'extreme overscorers'. Those lying below the regression line may be thought of as 'underscorers' again with those below the line at one standard error below the regression as 'extreme underscorers'. Thus the respondents from each of the seven areas can be divided into four categories. The respective percentages of respondents within each category are given in Table 5.8.

These figures are no more than first approximations to an estimate of the degree of over- and under-scoring in the seven areas. The proportions are obviously affected by the fact that an upper ceiling limit has been placed on the quantification of the attitude dimension. This means that it is virtually impossible for those families in high social classes to have attitude scores above the line of one standard error from the regression line. Nevertheless the pattern which emerges gives some clue to the nature of the types of attitudes in the areas. Three very general types of areas appear.

First, Deptford, Upper Hendon and Thorney Close are comprised of

TABLE 5.8. *Over- and Under-scorers*

Area	Overscorers (per cent)		Underscorers (per cent)		Total no. of families
	Extreme overscorers	Overscorers	Underscorers	Extreme underscorers	
a Deptford	11·8	35·3	23·5	29·4	17
b Upper Hendon	9·7	22·6	25·8	41·9	31
c Lower Hendon	37·5	37·5	12·5	12·5	16
d Thorney Close	9·4	21·9	43·7	25·0	32
e Bishopwearmouth	14·3	57·1	28·6	—	14
f Thornhill	13·3	46·7	33·3	6·7	30
g Alexandra Road	7·4	66·7	25·9	—	27

families which are predominantly underscorers. In each case, more than half of the families lie below the regression line of attitudes on social class, and very few families fall above one standard error above the regression line.

The second pattern is seen in Alexandra Road, Thornhill and Bishopwearmouth in which the great majority of respondents lie above the regression line. There are few 'underscorers' and no 'extreme underscorers'. The majority of respondents in each case fall in the 'overscorers' category with few 'extreme overscorers'. Thornhill differs in that it has larger percentages of its respondents lying in the 'underscorers' category and proportionally fewer in the 'overscorers'.

Lower Hendon is a third, and separate case unto itself. Over one-third of the families lie within the 'extreme overscorers' category; three-quarters are found above the regression line; a mere one-quarter lie below the regression line. The distribution of scores is therefore markedly skewed towards overscoring. Within this area, despite its relatively low social class, attitudes are therefore markedly favourable.

This grouping provides an interesting comment on the nature of the seven areas. If areal effects exist, they would appear to be operating to inhibit the development of aspiration for children's education in Deptford, Thorney Close and Upper Hendon and to facilitate the development of aspiration in Alexandra Road and, to a lesser degree, in Bishopwearmouth and Thornhill. The most dramatic effects appear to be in Lower Hendon where aspiration would appear to be greatly facilitated. Indeed, this area presents a quite startling case. The class of the respondents, as measured by the fathers' occupations, is only marginally above that in Deptford and Upper Hendon, the two other privately housed working-class areas. It was noted above, too, that in terms of the social class of the total population of the areas (as against the questionnaire populations), the three areas differed but little (see Table

5.1, above). Yet, in spite of this similarity to other working-class areas, Lower Hendon's respondents have attitude scores very similar to Alexandra Road, the upper middle class area.

Ideally, these differences in the attitude scores of the seven areas should be expressed in some more precise fashion. It has been pointed out above that simply to use the average class and average attitude scores of each area as the basis of comparison, has the disadvantage that there is not a 1 : 1 ratio between the two. A method which provides a more valid estimate of the degree of over- and under-scoring in each area is to use the average attitude scores in each class to calculate an 'expected' score for each area, which can then be compared with the *actual* score. As an example of the procedure adopted, the case of Upper Hendon might be considered. It can be seen from Table 5.4 that, grouping the families in terms of the father's occupation, Upper Hendon has four families in social class 1, one in social class 2, eighteen in social class 3, and nine in social class 4. The average attitude scores for the total survey population of each of these classes are as follows (Table 5.4): class 1 has an average score of 10·1, class 2 of 15·8, class 3 of 24·8, and class 4 of 24·9. To calculate the 'expected' attitude score of Upper Hendon, the numbers in each class are therefore multiplied by these average attitude scores, and summated to give a total of 499·7. This total, divided by the number of families considered (in this case 32), gives an average 'expected' score of 15·62 which compares with the area's *actual* average score of 11·60. The average 'expected' attitude score thus represents the score which the area would have had if its respondents' attitudes had been directly related to their social class. It is obvious that if the average 'expected' attitude score is above the average actual attitudes score then the respondents' attitudes within that area have been predominantly distributed below the average for their class group and the area has 'underscored'. Conversely, if the average 'expected' attitude score is below the average actual attitude score then the area has overscored.

Table 5.9 shows the results of these calculations for each of the seven areas with the families grouped in terms of their social class as determined by father's occupation (*a*), mother's work before marriage (*b*), and the combined class of father and mother (*c*). The results are shown graphically in Fig. 5.3. The figures become clear if one considers the basic premise that attitudes towards education are closely related to social class. If this relationship were perfect, one would anticipate that the average 'expected' attitude scores and the average actual attitude scores would be identical. The line at 100 per cent in Fig. 5.3 represents this case of absolute parity between the two. Distances from this line therefore represent a measure of the areal forces acting to distort a pattern based entirely on social class. Areas lying above 100 per cent

TABLE 5.9. *Comparison of actual and expected average attitude scores*

(a) Families grouped by father's class

Area	Average actual score	Average expected score	Actual as percentage of expected score
a Deptford	10·42	14·32	72·8
b Upper Hendon	11·60	15·62	74·3
c Lower Hendon	22·68	17·18	132·0
d Thorney Close	11·11	14·75	75·3
e Bishopwearmouth	22·93	19·28	118·9
f Thornhill	22·55	18·93	119·1
g Alexandra Road	25·55	22·33	114·4

(b) Families grouped by mother's class

Area	Average actual score	Average expected score	Actual as percentage of expected score
a Deptford	10·42	14·87	70·1
b Upper Hendon	11·60	15·44	75·1
c Lower Hendon	22·68	19·89	114·0
d Thorney Close	11·11	14·23	78·1
e Bishopwearmouth	22·93	19·81	115·7
f Thornhill	22·55	19·89	113·4
g Alexandra Road	25·55	21·44	119·1

(c) Families grouped by combined class score

Area	Average actual score	Average expected score	Actual as percentage of expected score
a Deptford	10·42	13·41	77·7
b Upper Hendon	11·60	15·26	76·0
c Lower Hendon	22·68	19·09	118·8
d Thorney Close	11·11	13·24	83·9
e Bishopwearmouth	22·93	20·32	112·8
f Thornhill	22·55	20·15	111·9
g Alexandra Road	25·55	24·03	106·3

show areal forces tending to increase the favourableness of attitudes towards education; those below show areal forces operating against this.

The marked differences between the actual and the 'expected' scores show that Lower Hendon has the greatest positive anomaly. Its tendency to over-score is thus again underlined. The three other areas with positive anomalies are Alexandra Road, Thornhill and Bishopwearmouth, each of which over-scores to about the same amount. The remaining three areas of Deptford, Upper Hendon and Thorney Close each underscore to approximately the same degree, with Thorney Close marginally closer to the 'expected' score than the other two areas.

It would therefore appear that there is an areal effect or an areal pattern readily discernible in these figures. It might be asked how consistent this pattern is. Does a particular area tend to have higher or lower attitude scores than its class would suggest only in certain social classes within the area? In other words, is it the effect of a minority group in each area which pulls

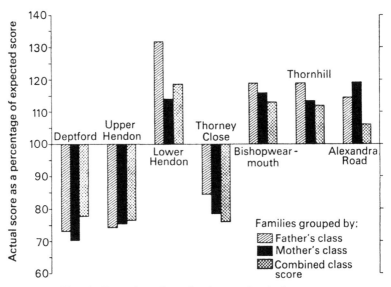

Fig. 5.3. Comparison of actual and expected attitude scores
Source : Field Survey

the average attitude score either above or below the expected average? To examine whether or not this is so, the respondents in each area must be subdivided into social classes so that it is possible to consider the classes within each area rather than each area as a whole. Attitude scores of respondents in similar classes but from different areas will then be directly comparable. A difficulty arises here since, as might be expected, the range of social classes is not equally or even fully represented in each area. Ideally, families in class 1 in Deptford should be compared with families in class 1 in each of the other six areas, and so on through each class and every area. But since there are no families in class 1 in Deptford, and other classes are absent in other areas, a full comparison of this sort cannot be made. This is partly the result of the necessarily small numbers involved in the particular sampling universe used here, but also, more importantly and unavoidably, it

TABLE 5.10. *Attitude scores holding area and social class constant*

(a) Families grouped by father's class

Area	Average attitude scores Class 1	Class 2	Class 3	Class 4	Number of families in each cell of table			
a Deptford	.	17·0	**11·1**	**8·8**	.	1	10	6
b Upper Hendon	19·5	29·0	**9·5**	12·1	4	1	18	9
c Lower Hendon	24·5	24·0	**25·1**	1·0	2	2	12	2
d Thorney Close	.	14·0	**12·4**	**8·4**	.	3	20	11
e Bishopwearmouth	28·0	29·0	**18·9**	19·0	4	2	7	1
f Thornhill	**21·9**	**25·9**	**22·0**	16·0	7	7	18	3
g Alexandra Road	**27·5**	**27·9**	**22·3**	11·0	13	11	7	1

(b) Families grouped by mother's class

Area	Average attitude scores Class 1	Class 2	Class 3	Class 4	Number of families in each cell of table			
a Deptford	.	7·3	.	12·4	.	6	.	11
b Upper Hendon	26·0	12·2	13·7	10·0	1	10	7	15
c Lower Hendon	.	**24·8**	16·0	29·0	.	13	3	1
d Thorney Close	.	**18·0**	14·3	8·5	.	9	3	23
e Bishopwearmouth	29·0	**23·6**	.	18·7	1	10	.	3
f Thornhill	.	**22·8**	28·0	**18·2**	.	28	1	5
g Alexandra Road	28·0	**25·9**	.	11·0	1	26	.	1

(c) Families grouped by combined class score

Area	Average attitude scores Class 2	Class 3	Class 4	Class 5	Class 6	Class 7	Class 8	Number of families in each cell of table						
a	.	.	.	1·2	14·0	**15·1**	4·7	.	.	.	3	4	7	3
b	26·0	24·0	16·0	12·5	10·1	11·0	10·2	1	1	1	8	7	8	5
c	.	24·5	29·0	23·4	29·0	14·5	.	.	2	1	10	1	2	.
d	.	.	25·0	17·4	14·3	10·6	15·2	.	.	1	5	7	12	8
e	29·0	27·7	29·0	19·0	.	18·5	19·0	1	3	2	5	.	2	1
f	.	**25·2**	**25·8**	21·1	20·4	19·0	.	.	5	6	16	5	2	.
g	.	**27·4**	**26·6**	21·8	.	.	11·0	.	11	10	5	.	.	1

NOTES. *i* bold figures indicate cells which include five and more families.

ii Total numbers for each area may differ since data on father's and mother's class could not be assigned to an appropriate category in all cases.

iii Tables are arranged so that higher social classes are to the left, lower classes are to the right.

results from the relatively high degree of social homogeneity within most of the areas.

Table 5.10 shows the attitude scores holding constant both the area and the social class of each respondent. The number of cases within each cell of the table is recorded and those cells which include more than five respondents have been differentiated. It can be seen that, in each of the three sets of figures, there is only one social class grouping in which there is adequate material to permit within-class comparisons between the areas to be made. For the families grouped in terms of father's occupation, this is class 3 (the skilled manual class); for the grouping based on mother's class, it is class 2 (the lower middle class); for the grouping based on combined class scores it is class score 5.

If these three class groups are first considered, it can be seen that people of similar class display very different attitudes towards education depending on the area in which they live. In all three tables the intra-*class* variation in average scores is greater than the intra-*area* variation. Thus, for example, Deptford has a range of attitude scores from 8·8 to 17·0 in the first table (*a*), whereas class 3 has a range from 9·5 to 25·1. It is also interesting to note that Lower Hendon actually has the highest average score of any of the areas in class 3 in.table (*a*); it is second highest in table (*b*); and it is highest in table (*c*). Again, the quite extraordinarily high degree of educational aspiration within the area is underlined.

The most important point to note about Table 5.10, however, is that each set of tables shows that within any one social class there is a consistently higher attitude score in Alexandra Road, Lower Hendon, Thornhill and Bishopwearmouth than is found in Deptford, Upper Hendon or Thorney Close. This difference is particularly marked amongst the central groups of the social classes, where there is the greatest degree of overlap of respondents in the same class in different areas, and where one might reasonably expect to find the greatest degree of variation in the average attitude scores since the people within these classes are in something of a marginal position between the working class and the middle class. This pattern of consistent areal differentiation in attitude scores holds true in every case within the tables where the number of respondents within any cell is over five. Even including those cases where the number of respondents is below five, in the whole of the three tables, out of a total of eighty cells in which there is at least one entry, there are only four cases in which the four areas of Alexandra Road, Lower Hendon, Thornhill, and Bishopwearmouth have not got higher average attitude scores than all of the three remaining areas. All four of these anomalous cases occur where there is only one case in the cell.

It would therefore appear that there is a marked areal effect apparent in the development of attitudes towards education as measured by the survey of parents whose sons were about to sit the 11-plus examination. This areal effect inhibits the development of favourable attitudes in certain areas and encourages it in others. Lower Hendon, the skilled working-class area, has been seen to exhibit the most interesting and the largest deviation from the pattern of scores which would be expected if social class alone determined what sort of attitudes prevailed. In the following chapter, these discrepancies are examined in more detail in an attempt to reach some conclusions about those forces at work within local areas which might be held responsible for the wide areal variation of intra-class attitude scores which has been noted here.

CHAPTER 6

EDUCATION AND NEIGHBOURHOOD

The previous chapter has shown that within Sunderland, the area in which an individual lives appears to bear a strong relationship to his attitudes towards education even when the class membership of the individual is held constant. Location can be considered to be exerting certain influences in the determination of the attitudes which are held, and this locational influence has been isolated by studying the individuals as part of groups in space rather than as isolated units. This locational aspect of the development of attitudes has not been fully pursued in the extensive literature on education. The plethora of books and articles on the subject of education almost invariably take as their starting point or frame of reference the class structure and the individual's placement within the class structure.[1] Such an approach tends to abstract the individual from the social setting in which his attitudes are largely formed. Having suggested that this milieu—both social and physical —plays a part in the formation of attitudes, this chapter will look more closely at the role of the neighbourhood, treating the questionnaire material from the geographical point of view of location rather than the socio-psychological approach of the case-study of particular individuals. Before considering what are the forces through which the neighbourhood might exert this influence, however, some reference will be made to work within this field which has some bearing on the results of the Sunderland survey.

In considering the role of the external environment on patterns of social behaviour and social attitudes, one has to tread carefully amongst the relevant research. While the effect of place and neighbourhood upon social structure has formed one of the mainstreams of social research, there has nevertheless been little comparative work done on the effects of different urban environments on the psychology, behaviour and attitudes of urban dwellers. Hatt and Reiss consider, for example, that while there has been much work done on the differences between the opinions and attitudes of urban and rural

---

[1] Even in instances where the physical or social background is a primary focus, invariably it is viewed at an individual or case-study level rather than as an areal or locational factor. See, for example, E. Frazer, *Home environment and school* (London, 1959), and W. J. Campbell, 'The influence of the home environment on educational progress in secondary schools', *Br. J. Educ. Psychol.* XXII (1952), 89–100.

inhabitants and of inhabitants of different-sized urban centres, 'there none the less are few studies of the extent to which urban living conditions these attitudes and opinions, since most of the correlates are demographic or stratification variables'.[1] Few field studies have attempted to follow up the lead of Plant by studying *intra*-urban variations in personality development.[2]

The mass of evidence on the effects of the external environment suggests that it does indeed play a role in the development of a wide variety of social phenomena. The difficulty lies in teasing out the complex relations between the variables involved, since simple correlations cannot provide convincing evidence in the absence of any theoretical rationale underlying the associations which are isolated. This is less serious in the case of established associations between housing and certain aspects of physical health where connections can be established with some degree of certainty. An association between poor housing and the incidence of infectious diseases or of respiratory diseases can be explained in terms of medical knowledge of the causation of such diseases. But one is on less certain ground in examining environmental connections with social traits such as delinquency, family relations, personality development or attitudes. Lauder, for example, has shown that while there is a simple correlation between juvenile delinquency and poor housing, when variables such as education levels are held constant the associations disappear, showing that the connection is not causal.[3] In other studies which attempt to isolate the effects of housing through comparisons of people in different types of housing, the difficulty again is one of controlling the distorting effects of intervening variables. Gans, for example, argues that, in making comparisons between families which have moved to new estates and those which are left in older areas, while differences in the degree of socialization may be found, it is not the new environment which causes these differences. Rather, the desire to socialize is one of the motivations underlying the move.[4]

Nevertheless, despite the complexities, the conclusions of the available research do suggest a *prima facie* case for the effect of the external environment. Schorr, after an admirably cautious review of the evidence, concludes that, while there is only a hint of the importance of such factors as the internal arrangement of housing, there is overwhelming evidence of the effects of *extremely* poor housing on the conditioning of behaviour and attitudes and that 'the placement of houses and flats in relation to one another and to the

[1] P. K. Hatt and A. J. Reiss (eds.), *Cities and society: the revised reader in urban sociology* (second edition, Glencoe, Illinois, 1957), p. 9. This emphasis on urban/rural differences applies much less to British than to American work.

[2] J. S. Plant, 'The personality and an urban area', in Hatt and Reiss (eds.), pp. 647–65.

[3] B. Lauder, *Towards an understanding of juvenile delinquency* (New York, 1954), p. 79.

[4] H. J. Gans, 'Planning and social life: friendship and neighbor relations in sub–communities', *J. Am. Inst. of Planners*, XXVII (1961), 134–40.

total city (downtown, suburban) clearly influences family and social relations'.[1] This distinction which Schorr draws between the effects of housing *per se*, and of the neighbourhood as a whole, is conceptually useful even though the two overlap to a very great extent. Various studies have concentrated on the effects of improved housing itself and have shown somewhat ambiguous evidence. While some suggest that improved aspiration results, others suggest that, while morale improves, only certain of the people involved actually respond to the improvement in housing conditions. Children in particular are likely to benefit from better housing even though such changes in adults may only be effected where the improved housing is accompanied by real changes in opportunity for the individuals concerned.[2]

Many of the changes in physical housing are closely related to changes in the whole complex of neighbourhood stimuli. Much of the research involving the role of neighbourhood has concentrated either on the analysis of communities in relatively impoverished and dilapidated areas or on comparisons of such areas with areas of rehoused population. Comparisons have been made between planned and unplanned neighbourhoods or between old and new neighbourhoods and frequently the comparisons have been made through before-and-after studies of the rehoused population. Bearing in mind the limitations of such comparisons, some moderately consistent conclusions have been drawn. The long-established neighbourhoods of working-class population have tended to be close-knit, with wide interactions between individuals. Patterns of extended family relations and wide networks of informal neighbour relations have been suggested as characteristic of such areas. With the dislocation involved in the move to new areas, many of the traditional social ties have been altered. In view of the fact that people who reside close to each other tend to become friends or to form social networks,[3] such changes in social patterns are not unexpected. The new juxtaposition of individuals combined with the different physical milieu of a new housing estate act as triggers which spark off changes in the life styles of the individuals and families concerned. Amongst the changes which have been noted, the most consistently reported is the tendency for the nuclear family to replace the extended family as the basic social unit.[4] There is some evi-

---

[1] A. L. Schorr, *Slums and social insecurity* (London, 1964).

[2] Cf. D. M. Wilner, R. P. Walkney *et al.*, *The housing environment and family life* (Baltimore, 1962), and M. Millspaugh and G. Breckenfeld, *The human side of urban renewal* (Baltimore, 1958). For effects on children see W. S. Jackson, 'Housing and pupil growth and development', *J. Educ. Sociol.* XXVIII (1955), 370–80.

[3] L. Festinger, S. Schachter and K. Back, *Social pressures in informal groups* (New York, 1950); W. H. Form, 'Stratification in low and middle income housing areas', *J. Social Issues*, VII (1951), 109–31; W. H. Whyte, Jr., *The organization man* (London, 1957).

[4] J. M. Mogey, *Family and neighbourhood* (London, 1956); M. Young and P. Willmott, *Family and kinship in East London* (London, 1957); A. G. Mitchell *et al.*, *Neighbourhood and community* (Liverpool, 1954).

218

dence, however, to suggest that such changes are merely temporary phenomena or, indeed, are the result of differences between the control population
and the rehoused population.[1]

Apart from such studies of actual relocation, there have been relatively
few studies of locational effects on social patterns, especially within Britain.
A notable exception is Bott's work on the respective roles of husbands and
wives within the urban family. Bott attempts to relate the different functional
roles of the two to variations in the local social environment. Indeed, she
makes the point that, within such studies, more attention should be given
to the ecological aspects of behavioural characteristics: 'A comparative study
of neighbourhoods in which ecological analysis was combined with analysis
of formal institutions and actual social relationships of families in the areas,
would fill an important gap in the theoretical and factual knowledge...there
is a great need of...a study that would analyse type of familial network in
relation to type of area.'[2]

In America, this need has partly been met by certain studies which have
concentrated on neighbourhood characteristics rather than individual
attributes. The results of this work suggest support for Park's early claim
that 'people living in natural areas of the same general type, and subject to the
same social conditions will display on the whole the same characteristics'. One
trait which has been rather extensively used as an index of patterns of social
behaviour is the degree of participation in formal associations, and it has
been demonstrated that in this at least, when personal social class or income
characteristics are held constant, there are variations in social behaviour
which are concentrated in different urban sub-areas.[3] Using this same
parameter Bell and Force have shown similar results: that different neighbourhoods can show very different patterns of participation in formal clubs
and societies.[4] The finding most pertinent to the present study is that when
their respondents' behaviour was treated as a personal variable rather than a
neighbourhood characteristic, the predominant neighbourhood patterns
tended to overcome the variations which might have been expected to arise
from the personal social variables. The neighbourhood, in other words,
appeared more important than the personal social characteristics of the
individuals in defining the individual's pattern of social behaviour. They
therefore conclude: 'at each of the educational levels, the men living in the

[1] Studying Dagenham some forty years after its establishment as an out-country London estate,
Willmott suggests that traditional patterns of behaviour tend to re-assert themselves. Cf. P.
Willmott, *The evolution of a community: a study of Dagenham after 40 years* (London, 1963).
[2] E. Bott, *Family and social network* (London, 1957), p. 227.
[3] S. Greer, 'Urbanism reconsidered: a comparative study of local areas in a metropolis', *Am.
Sociol. Rev.* XXI (1956), 19–25.
[4] W. Bell and M. T. Force, 'Urban neighborhood types and participation in formal associations',
Am. Sociol. Rev. XXI (1956), 25–34.

higher status economic neighbourhoods are more likely to be more frequent attenders (at the clubs and societies) than are the men living in the lower economic status neighbourhoods'.[1] The same finding resulted when occupation and income were held constant in place of education. Differences in participation in formal associations therefore still existed when comparing the high and low status areas, even when such measures of personal social and economic characteristics were controlled.

This is a most interesting finding since, albeit with a different dependent variable, it is exactly parallel to the findings of the Sunderland survey which were outlined in the previous chapter. When social class was held constant, attitude scores still varied consistently from area to area. The reasons which Bell and Force advance to explain their findings are also apposite. It is suggested, first, that the neighbourhood characteristics may be an index to the self-image of the individual, and second, that the neighbourhood itself might be a factor in the kinds of pressures which are brought to bear on the individual to participate in formal associations. To elaborate this, it might be suggested that, in the first case, the individual's area of residence may give clues to his own reference group. In the second case, the neighbourhood itself would be involved in determining the role expectations of the individual.

Here then are two kinds of way in which the neighbourhood may either reveal or may mould the type of attitudes which the individual holds. They can be thought of in terms of *reflecting* and *affecting* the individual's behaviour. While Bell and Force make no attempt to elaborate this theme and do not discuss the sort of forces which they may have had in mind as being those which the neighbourhood can exert, their dichotomy is useful in considering the findings of the Sunderland study with regard to educational attitudes.

The first explanation of the areal differential is one which can be suggested, but not easily substantiated. In thinking of the neighbourhood as a reflection of the self-image of the individual, one runs immediately into the problem of cause and effect. It is impossible to disentangle the complex Gordian knot of whether the individual living in an area comes to reflect the attitude and behaviour of that area, or whether he has moved to the area because its attitudinal and behavioural norms were close to his own. Suffice it to say that Sunderland, no less than the Boston of Walter Firey, has its own equivalent of Beacon Hill and that the sub-areas of the town can be ranked in terms of a scale of social desirability. Areas have their own symbolic content as much as a physical and economic one, as Firey took pains to emphasize.

The second explanation, that neighbourhoods affect rather than reflect the individual's attitudes, is one which can more easily be tested. What, it may

[1] *Ibid.* p. 31.

be asked, are the forces which could operate through the sub-areas of a town to influence the development of attitudes towards education and to explain the patterns which have been noted?

AREAL FORCES

The community ethos

Education in Britain is very much a preserve of the middle class. This fact remains even though changes since 1944 have done much to erode the middle-class composition of the grammar school.[1] Despite the greater number of children from the lower reaches of the social scale who now gain grammar school places and who progress to some form of higher education, the assimilation of the grammar school ethos seems to have been an uneasy one. Jackson and Marsden depressingly show how the change has been one in letter rather than in spirit.[2] There is still uneasiness and unfamiliarity in the face of the grammar school tradition. This is especially likely to be so in areas where few children gain grammar school places. There are areas within Sunderland which are very similar to Jackson and Marsden's 'Railway Street' area within which no children gained places at grammar schools. In the East End of Sunderland, fewer than 10 per cent of the children were successful in the 11-plus examinations in the two years 1962 and 1963. By contrast, the Alexandra Road area had pass rates of 80 per cent and over. These are the extreme cases within the town, but the seven areas selected for the question-naire survey showed almost as great a variation in their success rates, with Thorney Close having the lowest and Alexandra Road the highest. These differences form a vicious circle which serves only to widen the discrepancy between the areas, or to maintain the differentials. In the area which gains few grammar school places, the secondary modern school is the norm, the accepted channel through which children pass on their way to an early job; in the area with high pass rates, the grammar school assumes this role of accepted channel and leads to a very different process. Furthermore, the secondary modern school plays a part in the life of the working-class com-munity. From work done elsewhere and from the results of the survey in Sunderland, it would seem clear that the working-class sense of community has a geographical base, which is not found in the middle class area.[3] The

---

[1] See J. Floud, 'Education and social class in the welfare state', in A. V. Judges (ed.), *Looking forward in education* (London, 1955), esp. pp. 44 ff. See also H. T. Himmelweit, 'Social status and secondary education since the 1944 Act: some data for London', in D. V. Glass (ed.), *Social mobility in Britain* (London, 1954), pp. 141–59.

[2] B. Jackson and D. Marsden, *Education and the working class* (London, 1962).

[3] The working class 'territorial sense' is well distinguished from the 'selective and limited use of space' by the middle class in M. Fried and P. Gleicher, 'Some sources of residential satisfaction in an urban slum', *J. Am. Inst. Planners*, XXVII (1961), 305–15. The varying range of geographic

working class community sense depends partly on proximity which gives rise to face-to-face familiarity and common shared experience (although family ties may be important even where the family is geographically separated); whereas the middle class sense of community is based more on shared opinions and values which can be disseminated as easily by reading matter and other forms of communication as by common residence in a given area. The geographical immobility of the working class is perhaps a reason for its parochialism and conservatism. With this geographically based sense of community, the local secondary modern school, built within the local area, falls naturally into a neighbourhood role; a role which the grammar school, usually built outside the immediate area of residence, cannot as easily play. That the grammar school is built at a distance from the working class area is partly a reflection of the social class for which it caters and also of the small but widely scattered, non-local population which it serves. Such physical distance has, for the working class area, a connotation of foreignness which it has not necessarily got for the middle class area. Without labouring the theme, it would seem that the middle class parent can more easily accept the grammar school as 'their' school even though it is not built within their own neighbourhood. For the working class population, the fact of its physical distance merely reinforces the fact that the grammar school disseminates middle class standards anyway.

This reaction to the grammar school as necessarily foreign and the projection from this to a more general mistrust of education as such, was encountered time and again in interviews with respondents from Deptford, Upper Hendon and Thorney Close. In Upper Hendon, it was implicit in the acceptance of the local secondary school in Hendon as the normal and, in fact, the only possible educational path for children of the area. Many of the parents had themselves been educated at this same 'Board' school before it was divided into infant, junior and senior sections. They had spent the whole of their formal education within the single school, and the fact of the age streaming was either not understood fully, or regarded as irrelevant. The child's progress through the various stages of the same school was seen as inevitable and any other pattern was unthinkable for the majority of parents. Again, the mistrust of the grammar school was evidenced by frequent denigratory remarks made about them. Stories of boys who had possessions stolen from them in the local grammar school were capped with the comment, 'and that's supposed to be one of them good schools'. Or again, a revealing comment was made by one woman in Deptford who, when asked about further education, said 'Well, I don't know. If you mean Oxford or anything

horizons in the two classes is discussed in M. Stacey, *Tradition and change: a study of Banbury* (London, 1960), ch. 8.

like that, I don't think that sort of thing's for our class round here.' Another woman in Deptford told of her eldest son having passed the 11-plus examination and said, 'Mind, I didn't think he had a chance. He said he would pass, but I just laughed at him. I didn't dare to face the neighbours when he got through.' It is easy to read into such brief excerpts the sort of interpretation that one wishes, but in the context of the total interview in each case, they all pointed to the feeling of estrangement from the grammar school and a consequent reaction against or rejection of the system as a whole.

Willmott and Young suggest that one of the ways in which the working class can find a basis for self-respect is in the extolling of the worth of manual labour and the conscious isolation of working class areas from middle class influences.[1] The object is to form a one-class community isolated from, and therefore immune from the criticism of and the comparison with, other types of area and influence. To aspire to rise above one's born situation will therefore represent a crack in this solidarity, a breaking of the ranks of isolation. This obviously applies with especial force to educational aspiration since this is the main avenue of social mobility in our present society. This feeling of solidarity and the sense of security which it engenders will vary according to the specific context within which it is formed. Willmott and Young point to the differences between the working class districts of Bethnal Green and Woodford.[2] In Woodford, the working class is a minority element and is found to share many more of the behavioural and attitudinal traits of the adjacent middle class community than do the working class people of Bethnal Green who, since they live in so exclusively a one-class area, have much less in common with the middle class in terms of their attitudes and patterns of behaviour.

This is an excellent example of the kind of locational or ecological forces which the Sunderland survey has attempted to isolate. If the same argument is applied to Sunderland as a whole, one could argue that since it is predominantly a working class town, the attitudes of any one social class within the town may well be less favourable towards education than those of a similar class in a more middle class town. Certainly Floud's study of Middlesbrough and south-west Hertfordshire showed marked differences between the two areas, and Middlesbrough is somewhat similar to Sunderland, in terms of its class composition.[3] Likewise, on the micro scale, it can be seen that the more solidly working class an area is, the more its attitudes towards education might be expected to be inhibited in consequence of this introverted working-class group mentality. It is not without significance that the most close-knit

[1] P. Willmott and M. Young, *Family and class in a London suburb* (London, 1960), pp. 129–32.
[2] *Ibid.* pp. 78–81 and 129–32.
[3] J. Floud, A. H. Halsey and F. M. Martin, *Social class and educational opportunity* (London, 1956), esp. chs. 5 and 6.

of the seven survey areas is Deptford, which is more homogeneous in its working class composition and the length of residence within the area of its inhabitants, than any other of the areas.

This isolation of the working class areas and the stultifying effects that it can have on the development of attitudes towards education is exacerbated by the fact that the catchment areas of the various schools tend to follow the socio-economic pattern of the town. Upper Hendon, for example, is served by a distinct set of primary schools in the Hendon area which cater for a solidly working class area with no overlap into the middle class areas. This serves to reinforce the identification of the local working class area with these local schools and further bolsters the isolation of Hendon from the mainstream, or perhaps it would be truer to say the minority-stream, of the grammar school tradition. The 11-plus becomes a formality and the grammar school very much an alien intrusion in so introvert and private a world. Furthermore, the fact that all the children from Deptford, Upper Hendon and Thorney Close and, to a certain extent Lower Hendon too, tend to go to the same school which caters for a solidly working class intake, has its own independent effects on all the children who attend. They are all tarred with the same brush. As Wilson suggested from American data, a school's 'moral climate' affects the motivations of children by providing an ethos in which values are perceived. He found that, even when parental occupation and education are held constant, children at predominantly working class schools have lower aspirations than those at higher class schools: a conclusion which accords well with the Sunderland findings.[1]

Such differences within the primary schools of the survey areas can be shown to be marked. The questionnaire included questions concerning the relationship between school and home and on the school's attitude to homework. Information on the pass rates of each school are available in the records of the Education Offices. From these sources, it emerges that the schools which serve Deptford had pass rates at the 11-plus examination of 1962 of only 27·3 and 17·6 per cent; those serving Hendon had pass rates of 11·5, 14·9 and 0 per cent; that serving Thorney Close had a pass rate of 18·2 per cent. Lower Hendon had slightly better rates of 19·0 and 37·2 per cent in its two schools. Thornhill's school had a pass rate of 37·9 per cent; that serving the Alexandra Road area had a pass rate of as high as 53·1 per cent. Considering the extremes of Upper Hendon and Alexandra Road, it is obvious that a school with a pass rate of less than 15 per cent is less likely to gear its efforts to preparing its children for the 11-plus than is one with a pass rate of over one-half. In fact, the questionnaire showed that the primary school

[1] See A. B. Wilson, 'Residential segregation of social classes and aspirations of High School boys', *Am. Sociol. Rev.* XXIV (1959), 836–45.

serving the Alexandra Road area begins its preparation for the examination a year before it is due to be taken, and sets regular homework to boys of 9 and 10. It also holds regular Open Days at which the parents are invited to ask advice of the teachers. The schools of Upper Hendon, on the other hand, only set homework, and that rather sporadically and only to certain pupils, in the term immediately preceding the examination. Nor is there such specific and adequate provision made for parents to visit the schools. The questionnaire showed, in fact, that there are large differences in the likelihood and the frequency of parents visiting the schools in the two areas. This, in addition to the fact, as Mays points out for Liverpool, that the middle class parents are more likely to know the teachers in any case.[1]

Thus the first way in which the neighbourhood can have an independent effect in moulding or affecting the development of parental attitudes towards education is through this conflict between the community ethos and the educational ethos associated with the grammar school and all things academic. This conflict has been widely recognized and commented upon in the body of sociological material relating to education. With a primary interest in ecological patterns, however, we might go further to ask whether there are any factors within each neighbourhood itself which can heighten or strengthen this conflict. Two main factors appear to emerge from the results of the questionnaire: first is the effect of crowded living conditions, and second the influence of the number of children in an area. Both factors assume some importance because of the particular type of personality which is demanded by education and social mobility.

Education is essentially an individual exercise. In so far as it is elite-orientated and selective, it involves the acceptance of a minority of the population at the expense of the rejection of the remainder. To this extent, education is an anti-social process in that the individual is specifically encouraged to prove his own worth and individuality in order to further his own ends. The extreme archetype is the lonely academic and the ivory tower. But the trend can equally be seen in the atmosphere of heightened competitiveness which characterizes the successful grammar school. The individual who is exposed to pressures making for the development of greater individuality rather than adherence to group sentiments is thus probably more likely to succeed in the educational field.

Overcrowding

But overcrowding and the number of children in an area have relevance here. If, first, the effects of overcrowding are considered, the direct effects

[1] J. B. Mays, *Education and the urban child* (Liverpool, 1962), ch. 6.

on educational progress of living in overcrowded accommodation can readily be seen: the lack of space to do homework, the greater likelihood of illness and therefore of more days of absence from school and the debilitating effects of apathy and sluggishness. But what are the psychological effects of crowding in relation to personality development? Plant gave an early useful start to the investigation of this question, even though it was a start which has not been pursued in detail since. He attempts a thumbnail sketch of the effects of a crowded environment on the socialization and mental development of children. This throws up interesting conclusions about the sense of insecurity, the sense of inferiority and the lack of development of an independent personality which often arises. Life in a populous neighbourhood can breed a lack of self-sufficiency. Plant comments, 'It is as though they felt incomplete— without the necessary supports for the personality. It seems that persistent and constant crowding from early life destroys the sense of individuality— which without doubt is fostered by opportunities for privacy.'[1] He goes on to argue that this lack of self-sufficiency is expressed by an inability to be alone, so that 'the search is for games, for work where many others are close by'. The relevance of this point is evident since, if the individual is unable to support himself, he is less well suited for the educational process. If the ideals of the community are group ideals which revolve around the support of, and compliance with the group, an emphasis on formal education and the mobility which this implies becomes an alien intrusion: on the other hand, if the community ideals are personal, individualistic ideals, then education is a natural channel through which they can find expression.

The dichotomy between the working-class and the middle-class viewpoints in this respect is drawn by Goldthorpe and Lockwood, who designate the middle class position as 'individualistic' and the working class position as 'collectivistic'.[2] They argue that the 'public sociability' of the working class areas is maintained and consolidated by homogeneity of social composition and by stability of residence.[3] In terms of personality development, density or crowding would seem to play a part. The effects of density and this 'public sociability' are intensified by the fact that success in the close-knit community is measured in terms of success within the group. The attributes which are extolled become shrewdness rather than the kind of intelligence which is measured in terms of the I.Q. test. This is the same sort of argument, although couched within a more analytical framework, as Davis' criticism of intelligence tests because of their emphasis on verbal rather than practical

---

[1] Plant, in Hatt and Reiss (eds.), *Cities and society*, p. 652.
[2] J. H. Goldthorpe and D. Lockwood, 'Affluence and the British class structure', *Sociol. Rev.* XI (1963), p. 146.
[3] *Idem*, 'The manual worker: affluence, aspirations and assimilation', paper presented to the Annual Meeting of the British Sociological Association, 1962.

content.[1] Plant illustrates this different assessment of achievement by 'the frequent lack of correlation between "intelligence"...and success in ordinary street play. The school teacher marvels—and is often irritated—over children's preferences in their play life and games for those who far from shine in the classroom. (This has been perhaps the most persistent and jarring challenge she has met to her faith in the relationship of what she teaches to the real problems of life.)'[2]

Young and Willmott also make the point, regarding the different assessment of personality in the working class and middle class areas, that the assessment of the individual in the working class area is not made in terms of a single factor such as intelligence, but rather the individual is assessed 'in the round'. One man is respected because he is good at darts, another because he is a good story-teller, another because he is a good worker, and so forth.[3] In such circumstances an elaborate status hierarchy is impossible and individual differences tend to be submerged.

The urban child in certain environments can thus become ill-suited to the personality demands made on him by education, and in response to this, and partly as a result of their own personality development, the parents' attitudes towards education will be more hostile, or more apathetic. Such a conclusion must apply especially to the areas of Deptford and Upper Hendon in which, particularly in the latter, there is great overcrowding, and in both of which the well-established network of family kinskip adds cohesion to an already close-knit society. An unconscious response to this non-alignment of the aims and attitudes of these areas, and those of the grammar school and education as a whole, is seen in the fact that it was in these areas that the highest percentages of respondents agreed with Question 30 in the questionnaire which refers to the vague and unpractical nature of education.

Age structure

The second factor which works in an ecological manner and affects the personality development of the children of sub-areas of the town, is the number of children living within the area. As was noted above, the age pyramids of the seven areas show considerable differences in this respect (see above, pp. 195–7 and Table 5.2). Table 5.2 shows that it is in Thorney Close, Upper Hendon and Deptford that the greatest percentages of children are found. Thorney Close has nearly 20 per cent of its population aged under 10, and over 25 per cent aged 10 to 19. Upper Hendon has 22·6 per cent

[1] A. Davis, *Social class influences upon learning* (Cambridge, Mass., 1948).
[2] Plant, *op. cit.* p. 652.
[3] Young and Willmott, *Family and kinship*, pp. 133–5.

aged 0 to 9. Deptford has 16·6 per cent aged 0 to 9. Lower Hendon has rather fewer with 14·5 per cent aged 0 to 9. The middle class area of Alexandra Road has large numbers of children (15·7 per cent aged 0 to 9) but, as is argued below, this has very different implications. Thornhill and Bishopwearmouth have the lowest percentages of children in their populations.

These bald percentage figures have to be qualified since it is essential to take into account the amount of contact that children in an area are allowed to have with their peers before any conclusions can be drawn from the percentage figures. From the replies given to the questionnaire survey, it appears that Lower Hendon is an area in which children are specifically discouraged from playing in the streets and thus in which the opportunity for the children to mix with and grow up in a child-dominated world is effectively reduced. It is an area much more akin to the middle class areas of Alexandra Road and Thornhill in that children tend to spend their free time in some form of supervised activity. There is a larger membership of Boy Scout organizations, and of clubs and societies. Also such activities as music lessons or speech training lessons were mentioned in these areas, whereas none of the respondents in Deptford, Upper Hendon or Thorney Close sent their sons to such extra-school activities. To this extent, the effects of the smaller number of children in such areas are heightened by the fact that children's leisure activity tends to be controlled more by adults and takes place to a greater extent in the presence of adult society.

The importance of this factor of street play is that it again tends to work against the development of self-sufficiency in the individual in a similar way to the effects that crowding has on the total society. Obviously the element of crowding will reinforce this tendency for child street play in certain areas since, where there is overcrowding, the home offers less potential and less attraction as an area of play. But if street play, which is largely an areal force resulting from an area's age structure and its housing conditions, inhibits the development of a personality with the ability to be, and to work, by itself, it has further implications too. The first is that since play assumes so important a part in the lives of the children, success within the play group will assume as great an importance as, or a greater importance than, success at school. This is the point which was made above with reference to the different means of assessing excellence in areas of different status. The hierarchy of Whyte's street corner gang was based on very different considerations from those which make for success at school.[1] Secondly, the street gang or the play group forms an important channel of socialization for the child who lives in an area with many children and who fraternizes with its peers. The vocabulary and the thought forms to which such children are

[1] W. F. Whyte, *Street corner society* (Chicago, 1943).

exposed will be very different from those of the child who is brought up in a more adult society. Davis has drawn attention to the social class influence upon learning and points out the deleterious effects, on the educational opportunities of the child from the lower social ranks, of the linguistic bias in educational tests and in education generally.[1] Socialization in the child-centred world of the play group or street gang is likely to provide a training and an attitude even less well suited and adjusted to the tenets of formal schooling than the straightforward working-class environment about which Davis is concerned.

To reiterate the factors which have been isolated as the forces which the neighbourhood brings to bear to mould and condition the kind of attitudes which will develop towards education, it can be said that the close-knit working class community area will tend to produce a personality type which is unsuited to, and in consequence hostile towards, education. This personality type is one which lacks self-sufficiency and so depends on the security of the wider community for support. The development of this collectivist person-ality is determined first by the need for such areas to draw apart from middle class influences and secondly by the influence of certain social and physical characteristics, especially overcrowding and the age structure of the area, which exert influences on the development of the personality by a variety of means. These factors are further reinforced, as far as education is concerned, by the fact that the schools which cater for such areas tend to be as homo-geneous in the social composition of their intake, because of the ecological segregation of the areas which they serve.

INTEGRATION WITHIN THE NEIGHBOURHOOD

But having shown how certain types of area can develop goals and *mores* which are in conflict with the aims of English education, and which can therefore inhibit the development of attitudes favourable to education, it is essential to consider the degree to which the individual respondents of the survey were integrated into their neighbourhoods. This will obviously affect the degree to which such individuals tend to conform to the *mores* of the neighbourhood round about them. It has been recorded briefly how parents can counteract the effects of living in an area in which there are a larger number of children by forbidding their own children to play with or associate

[1] 'The lower socio-economic groups have a different language culture than the higher groups... their mental capacity cannot be tested by asking them to define such words as "ambiguous" or "illumination" or to understand the meaning of such phrases as "is to" in the statement: loud *is to* sound as bright *is to* what?... These questions test only one's facility and training in middle-class linguistic culture' (Davis, *Social class influences*, pp. 82–3). See also B. Bernstein, 'Some sociological determinants of perception', *Br. J. Sociol.* IX (1958), 159–74, and *idem*, 'Language and social class', *ibid.* XI (1960), 271–6.

with the children of the neighbourhood. In the same way they can isolate themselves from the effects of the particular ethos of the area in which they live by cutting themselves off from their neighbours. In a close-knit area, unless individuals make some effort to isolate themselves from their neighbours, they will tend to conform to a greater or lesser degree to the *mores* and ethos of the area in which they live. Where the prevailing *mores* of an area is in conflict with a particular goal—in this case the development of aspiration for the child—close attachment to or identification with the neighbourhood will tend to repress the development of aspiration for the goal, in this case because the personality requirements of education are in conflict with those of the neighbourhood. The development of such aspiration can only be achieved in three types of social situation: within an area which approves of the goal; within an area in which there is little participation between neighbours and so lacks the power of social control and, in this way, is 'anomic'; or within an area which does not approve of the goal, but in which, by complete and often painful isolation, the aspirant individual can exist. With the exception of the first type of situation, it can be seen that the desire for the goal involves a greater or lesser degree of isolation from the neighbourhood.

Turning to the seven study areas, the relevance and validity of this argument can be tested within Sunderland. The degree of isolation of individuals was measured in terms of three categories: those who were integrated into the neighbourhood; those were were partly isolated; and those who were more or less completely isolated. These categories were based on the responses to Questions 37, 38 and 39. It was felt that a non-quantitative approach to this question was as valid as, if not more valid than a purely quantitative one. It would have been possible to ask how often individuals visited neighbours, how many they spoke to on an average each day, how many they knew by name, or to use some more quantitative measure of this sort. It was felt, however, that since the patterns of social contact vary so greatly from one area to another—for one type of area it is common for people to invite neighbours into their houses, whereas in another this is not the behavioural norm even though this does not necessarily imply that the area is any less friendly—it was considered better simply to ask the respondents to give their own subjective assessment of how well they were integrated within the area and, in a general way, how much contact they had with their neighbours and how much contact their neighbours had with each other. In fact, this proved to be a fruitful approach since it led to the respondents making direct comparisons between themselves and the other people of the area and thus gave a good idea not only of how much contact they actually had with other people, but also of whether or not it was felt that they differed in this from their neighbours. In a very subjective way it therefore gave a measure of the degree to which they identified with and conformed to the area in which they lived. The cases of extreme isolation were readily discernible since the individual families concerned were very aware of their isolation (whether it was an isolation that they chose for themselves or one that was forced on them by the people round about), and the respondents were very anxious to talk of this—either disapproving of their neighbours' conduct and standing aloof from it, or else bearing resentment towards them on the grounds of their snobbishness. The cases in which families were defined as partly isolated were more difficult to distinguish but the category generally embraces those who saw little or nothing of their neighbours and felt no particular desire or need to see more of them. Those families which were regarded as well integrated into the neighbourhood knew most of the people living close to them, had a good deal of contact with them, and more or less completely identified themselves with their neighbourhood.

TABLE 6.1. *Degree of integration within the neighbourhood*

		Number of families			Percentage of families		
	Area	Well integrated	Moderately isolated	Very isolated	Well integrated	Moderately isolated	Very isolated
a	Deptford	9	3	5	52·9	17·6	29·4
b	Upper Hendon	20	5	9	58·8	14·7	26·5
c	Lower Hendon	2	8	8	11·1	44·4	44·4
d	Thorney Close	14	17	5	38·9	47·2	13·9
e	Bishopwearmouth	3	7	4	21·4	50·0	28·6
f	Thornhill	13	20	4	35·1	54·1	10·8
g	Alexandra Road	21	4	7	65·6	12·5	21·9

SOURCE. Field Survey.

Table 6.1 shows the results of this analysis in the seven study areas. Deptford, Upper Hendon and Alexandra Road are very similar in that they each show large percentages of the respondents who were well integrated into the areas. In the case of Deptford and Upper Hendon, this is the result of the close-knit working–class society which has been discussed above. It is also intensified by the tendency for most of the families to be surrounded by large numbers of relatives. The fact of this close social contact between people within each of these two areas accords well with their pattern of consistent underscoring in terms of their attitudes towards education. It would indeed seem that in these areas the neighbourhood community acts as a constraining influence upon the development of attitudes of aspiration for the education of children. Looking at the two areas in more detail, this assertion can better be supported.

It has been argued above that in such close-knit areas, attitudes favourable towards education can only develop if the individual cuts himself off from the people around him. In Deptford, Table 6.1 shows that there is a minority of isolated families. There seem, in fact, to be three types of position with respect to the degree of integration within the neighbourhood. The vast majority of the families are well integrated. This involves not only a complex set of well-established friendships and acquaintanceships within the area, but also a well-developed network of relations. Individuals are surrounded by a cloud of witnesses. This helps to explain the high proportion of underscorers within the area. However, a few families within the area are partly or wholly isolated. Two types of isolates can be visualized. The first is the isolated family which is rejected by the neighbourhood, of which only a single example was found within the sample. As the mother involved commented: 'Everybody here knows everybody else. We don't bother with them. We don't come up to their standards! They're just snobbish though. We come from a district that's friendly, but this lot round here wouldn't give you houseroom. Give me the pitfolks back again.' This attitude differed radically from that of the vast majority of the respondents within Deptford who, by comparison, commented approvingly on the close network of social relations within the area.

The second type of isolated family is very different from the first in so far as the family itself rejects the people in the neighbourhood rather than being rejected by the neighbourhood. Whereas the first type of family may identify with the people around, although it is rejected or at least feels itself to be rejected by these people, this second type is motivated by a different mechanism. Its conscious and self-imposed isolation is the result of such families not wishing to identify with the standards of their neighbours: its reference group, in other words, is socially higher than the local group. There are five families which illustrate this second type of isolate, showing a rejection of the people round about them within Deptford.

If the average attitude scores of these three types of family (the integrated, the rejected and the rejecter) are compared, the integrated families have an average score of 7·7, those who are partly or wholly isolated from the area by their own choosing through their rejection of the neighbourhood have an average score of 13·2, the one family which feels itself rejected by the neighbourhood has a score of 1·0. While the numbers involved are admittedly very small, the pattern of scores would tend to substantiate the hypothesis of the inhibiting areal effects of such neighbourhoods. The one family which is isolated by the neighbourhood does not represent a conscious self-imposed cutting-off from the neighbourhood and thus aspiration would not be expected. Those families which have developed a self-imposed isolation have done so because of their rejection of the standards of the local community. They have made an attempt to break out of the constraints of neighbourhood. Their higher attitude scores reflect this. Indeed all the scores of Deptford which fall above the regression line of Table 5.6 (see above, pp. 206–9) and which thus represent overscoring families, are families which were either highly or moderately isolated within the area, thus supporting the hypothesis of the restraining role of the neighbourhood within such working class areas.

The pattern of scores and integration within the neighbourhood of Upper Hendon is an even more dramatic vindication of the hypothesis. Here the tenemented conditions and the greater degree of overcrowding than at Deptford might be expected to produce even stronger neighbourhood effects. Living so closely on top of one another, and often within the same partly subdivided house, families show a certain degree of suspicion of their neighbours and a desire to keep apart from them, but the greater force is the cohesion produced by the dependence on neighbours for support, help and friendship. Families are caught between two opposing pulls: on the one hand the pressures of the neighbourhood which lead to a close-knit society with neighbours in intimate physical and emotional contact; on the other hand a desire for some measure of privacy and of protection from the exposure of all the facets of their lives to the criticism and gossip of the neighbours. The pattern accords well with Plant's picture of the crowded environment where individuals are unable to draw apart and develop the attributes of self-sufficiency and independence. As in Deptford, there are in Upper Hendon a certain number of families which have isolated themselves from the neighbourhood. With much greater crowding and physical decay in the area, these rejecter families form quite striking exceptions within the setting in which they exist. Set in the midst of tenemented squalor are a few houses which are carefully painted and cared for. In some cases, these houses are not subdivided and the people tend to be of higher class than their neighbours. This is not always the case. In some instances, the families are living, like those around them, in tenemented buildings which lack the possibilities of easy privacy and yet these isolated individual families manage to seal themselves off from the neighbourhood and to maintain some greater degree of 'respectability' in their homes. While people of higher social class than that prevailing round about them tend to be more or less isolated, there are also others who are not necessarily of higher social class who are also isolated. The average attitude scores of the three groups, delineated in Table 6.1, show that the families who are integrated into the area have an average score of 7·8, those who are partly isolated score 12·1, while those families which are extremely isolated and have cut themselves off almost completely from the area have an average score of 12·1. Furthermore, all three of the families shown in Table 5·6 as 'extreme overscorers' are isolated from the area.

Both Deptford and Upper Hendon therefore appear to substantiate the hypothesis outlined earlier. The attitude scores of the council estate of Thorney Close can likewise be seen in terms of the same types of neighbourhood forces. Indeed, it is interesting to note that, despite the improvement in housing within the council estate, attitudes to education are virtually identical to the other two working class areas.

The fourth working class area of Lower Hendon also confirms the hypothesis but in a radically different fashion. If the other working class areas are cases of neighbourhoods which are close-knit and inhibit the development of attitudes favourable to education, except for those individuals who isolate themselves from the areal influence, Lower Hendon is an example of an area which is individualistic, not collectivistic, in that neighbourhood participation and involvement is kept to as low a level as possible so that explicitly transmitted group values do not develop. The families within the survey population appear to be undergoing a change in their reference group. Goldthorpe and Lockwood would place them in their category of 'privatized worker' in which the families have a working class identification but with isolated rather than integrated relationships.[1] Table 6.1 shows that of the seven areas, Lower Hendon had the smallest percentage of well integrated families, and the attitudes towards the neighbours which were expressed were ones of consciously trying to reduce group pressures within the area. Each family's goal is to achieve its own ends. Whether or not this is at the expense of the group is immaterial. Such an individualistic ethos positively encourages aspiration for the education of the children within the area, and the remarkably high attitude scores shown in the previous chapter for this area are a reflection of this aspiration which the lack of public sociability of the area permits.

Within the working class areas, it therefore appears that attitudes to education can be strongly affected by the social structuring of neighbourhoods and the pressures which are exerted upon the individual to maintain his allegiance to the locally based group. Sociability within such areas thus implies a dampening-down of individual aspiration. Within the middle class areas, however, sociability has very different connotations. As Table 6.1 shows, the middle class area of Alexandra Road showed the highest level of integration of any of the survey areas. The question arises, why should this integration not repress aspiration in middle class areas as it does in the working class neighbourhoods?

As has been argued above, the middle class view of and use of space is very different from that of the working class. Contacts and standards are not

[1] Goldthorpe and Lockwood,'Affluence and the British class structure',*Sociol. Rev.* XI (1963), 150.

234

restricted to the local area, but derive from a much wider geographical orbit. The standards which are internalized need not be taken from a group which is essentially local, but are disseminated through reading matter and other mass media. This makes the group, of which any one middle class area is but a part, much more national than local in its affiliations. Its ties are not inextricably bound to the local area. This can be seen both in the type and the geographical range of the social contacts which prevail within middle class areas. The contacts which are made in middle class areas differ in quality from those in the working class areas. There is a large body of research material which shows how much greater is the degree of participation in formal associations within middle class areas. Stacey, for example, stresses the importance of formal associations for the middle classes and considers that 'perhaps the single most striking finding of the study of neighbour relations...(is) that the quality and quantity of these relations change remarkably from one social class to another'.[1] Working class contacts, on the other hand, are much more likely to be informal in nature. Thus, while sociability may be widespread within the middle class areas, it would appear to make much less stringent demands upon the *total* actor involved. Individuals can therefore maintain contact with their neighbours yet not involve themselves fully in this contact since they retain an area of privacy, of security and of support in the home and in the family.

Secondly, the much greater geographical spread of group connections can be inferred from the distribution of birthplaces of the middle class survey respondents. Within the working class areas, the range of birthplaces was tightly limited to a small ambit within County Durham, whereas the middle class areas had a broad spread of birthplaces throughout the country as a whole. The percentages of parents born within Sunderland itself for example was as follows: Deptford, 77 per cent; Upper Hendon, 84 per cent; Lower Hendon, 83 per cent; Thorney Close, 86 per cent; Bishopwearmouth, 57 per cent; Thornhill, 74 per cent and Alexandra Road, 50 per cent. The working class world is much more strictly limited by geographical proximity, and these figures serve to underline the less potent force of immediate neighbourhood within the middle class areas.

It is such qualitative differences between working and middle class individuals which can explain the coexistence of individual aspiration and sociability within middle class areas. Since affiliations are not primarily to a local group but to ideas and standards which transcend the local area, individuals can strive for self-advancement without this representing a rejection of group standards. This fundamental difference between the working class world and the middle class world has been clarified by sociologists in the

---

[1] M. Stacey, *Tradition and change: a study of Banbury* (London, 1960), p. 115.

different models of society which the two classes hold. The distinction between the collectivistic working class and the individualistic middle class relates closely to the dichotomy which has been drawn between the 'power' and 'prestige' models of the class structure.[1] The power model, suggested as being held by the working class, views society as being sharply divided into two contending camps: the distinction between 'we' and 'they'. The prestige model of the middle class, on the other hand, views society as a hierarchy of positions, a ladder of prestige whose rungs are climbed largely through the aid of formal education. These two models obviously have relevance to the development of attitudes to education. The power model bolsters the development of the close-knit community which views education as a potential threat to the solidarity of the working class ('we') area. The prestige model, by comparison, will tend to engender an individualistic and aspirant approach for which formal education primarily caters.

It is because of this distinction that the middle class areas have to be viewed within a different frame of reference from the working class areas. What the analysis of attitudes within the working class area has attempted to show is that the type of neighbourhood forces which have been outlined can have independent stultifying effects on the development of parental aspiration, and that only within certain types of area, or in certain conditions, such as the isolation of families living in unfavourable areas, can aspiration be developed fully.

To say that educational aspiration is correlated with social class adds to an understanding of the forces which have created and which maintain these differences, but as Bott says, in studying the roles of mother and father within the normal family unit, 'I do not think it would be illuminating to attribute differences in conjugal segregation to social class differences even if the correlations were high. This is only the first stage of the analysis...It is essential to push the analysis further, to find out what factors in social class are relevant to conjugal segregation and how they actually produce an effect on the internal role structure of families.'[2] This, within the sphere of attitudes to education, was the point at which this analysis began, and in discussing the question the chosen approach has been essentially ecological. To explain the inhibition of favourable attitudes in working class areas a number of neighbourhood characteristics have been suggested as possible forces which militate against the likelihood of the emergence of a personality or a social structure which can accommodate a desire for education within its social goals.

[1] Cf. E. Bott, 'The concept of a class as a reference group', *Human Relations*, III (1954), 259–85, who suggests four types of commonly held models of society, which are compounded of varieties of the basic power/prestige dichotomy. Goldthorpe and Lockwood (1963) use these basic concepts in their work on *embourgeoisement*. [2] E. Bott, *Family and social network*, pp. 218–19.

PART FOUR

CONCLUSION

CHAPTER 7

DISCUSSION

While the framework of this book has been set within the context of urban social geography in general, the fact that most of the substantive work has related to a single town obviously imposes limits to the generalizations that can be made. No single town can be regarded as a perfect paradigm of British towns, let alone towns in general. Indeed, in view of the fact that the profusion of urban classification studies has not produced any scheme which has been shown to be generally valid, it is difficult to think of a single town as representative of any class of towns such as 'manufacturing', 'transportation centres', or 'mining towns'. Some of the peculiar features of Sunderland were examined in chapter 3, where it was shown that the town's long history of poverty and its dependence on a heavy-industry base have left it with social, housing and demographic characteristics which differ markedly from many other British towns. It is best to consider these features as intervening variables which might distort or alter certain aspects of the patterning of traits within the town. Nevertheless, the Sunderland results can be used, in the light of evidence from other urban areas, to make some tentative steps towards the formulation of possible lines of urban theory.

In looking at the town, one of the main interests has been the ecological structuring of its residential areas. It has been shown that there is a body of historical data contained in old rating records and other primary documentary sources which permits the reconstruction of plausible patterns of residential characteristics. It so happens that in Britain, unlike America, most of the interest in the historical processes of urban growth has been shown by economic historians. Yet the factors which controlled growth and internal differentiation in earlier periods can provide illuminating insights into certain of the current processes of urban development and, if useful and relevant study of urban areas in earlier time periods is to be undertaken, geographers have a great deal to learn from the economic historian whose ingenuity and imaginativeness in making full use of documentary evidence is an example worth following. In Sunderland, as in many other British towns, the nineteenth century is a period for which full use has not been made of primary sources of data, yet much greater use could be made of such material as rating valuations, where such exist, in reconstructing past geographies of the urban area. Such studies could help to fill some of the important lacunae in our knowledge. In Sunderland the evidence showed that the classical models of urban growth have a good deal of relevance in gaining an understanding of the evolution of residential areas, but that other factors such as the importance of industrial development have to be given greater weight in the analysis of urban structure. In the twentieth century, by comparison, the development of council housing and of town planning and the effects of a variety of social changes have had profound effects upon the urban scene and have largely invalidated many of the bases on which the classical models have been built. In their place, a more pragmatic manipulation of urban data by the multivariate technique of component analysis was used to reveal the essential structuring of a large number of characteristics within Sunderland. As was emphasized in the second chapter, component analysis appears to offer interesting prospects as a basic tool in urban analysis. It provides a means of handling a multiplicity of essential data in a parsimonious fashion and can thus identify the basic patterns of covariance of variables and provide a simplified areal breakdown of an urban area in terms of the evidence of a large number of sets of material. The few studies which have used the technique to analyse urban sub-areas and urban structure have shown a number of points of similarity which encourage one to be more sanguine about the possibility of teasing out some of the fundamental complexes of urban society. Already the first three roughly comparable studies of Merseyside, Hampshire and Sunderland have shown similar amounts of explanation of the total variability and, more interestingly, have shown that certain of the components are rather similarly constructed. The density of persons per room, for example, appears to be one of the most fundamental indexes of urban spatial differentiation and one which is closely related to a very wide range of social and demographic characteristics. While any evaluation of comparability depends upon further studies being undertaken with a view to testing the reliability and validity of results obtained in particular areas, the results so far achieved have illustrated the value of the technique and hold out prospects of a far more powerful understanding of urban structure.

In Sunderland the results of the component analysis showed that by collapsing the full set of data a number of basic complexes emerged which were related to social class characteristics, to two aspects of housing characteristics and to age structure. The age structure characteristics were shown to be of great importance and especially useful in demarcating differing types of council housing areas. It was on the basis of such clusterings that aggregations of social areas were formed within the town, and amongst those which clearly emerged were middle class areas, stable working class areas, subdivided working class areas of the twilight zone, rooming-house areas and council housing areas distinguished in terms of their age composition. Within this framework of social areas, the consequences and effects of certain environmental influences were then examined in relation to attitudes to education, and it was shown that the area of residence had strong associations with prevailing attitudes even when the personal variable of occupation was held constant. To this extent it was suggested that spatial effects were at work in influencing the formation of these attitudes and that, of the possible effects which should be considered, the age composition and social integration of the areas were of some importance in addition to the more obvious effects of the physical nature of the areas and the degree of crowding within them.

In trying to contribute to the formulation of urban theory, three main conclusions from the Sunderland study might be emphasized. First is the importance of the local area as a facet of the individual's social world. Despite the loosening of the individual's ties to his locality, the study of social differentiation within the town and of the analysis of attitudes to education within the context of such differentiated areas suggest that the urban sub-area is of great importance, whether as a reflection of the individual's self-placement in society or as an active agent in the formation of his beliefs and attitudes. Much of the existing theory of urbanism has painted its canvas on a much broader scale, treating the whole town as its basic unit, and couched in terms of dichotomies between rural and urban which are suggested as having consequences for the nature of social relationships and social networks. Tönnies' *Gemeinschaft* and *Gesellschaft*, Maine's 'status' and 'contract', Durkheim's 'mechanical' and 'organic solidarity' and many other such dichotomies are basic to sociological concepts of urbanism and they gave rise to the continua suggested by Redfield and Wirth which drew heavily upon these polarities. Redfield and Wirth both saw in urbanism an all-embracing trend to the break-up of social stability with the replacement of primary by secondary relationships and the increase of secularization and disorganization.[1] The change of heart which has been shown by social scientists and anthropologists towards the Redfield–Wirth argument reflects the over-simplification which such dichotomies introduced and also the accumulating evidence of a large body of empirical research which has contradicted their ideas of urbanism. The statements of faith which characterize much of Wirth's writing on the effects of size, density and heterogeneity have partly been replaced by more soundly based research demonstrating the variety of social behaviour and social ties not only amongst more primitive 'folk'

[1] R. Redfield, *The folk culture of Yucatan* (Chicago, 1941); L. Wirth, 'Urbanism as a way of life', *Am. J. Sociol.* XLIV (1938), 1-24.

societies, but also as between different groups within the urban areas. Studies in Bethnal Green or in Boston's West End, for example, have shown the importance of kinship networks and face-to-face relationships within parts of the urban area and Lewis, amongst the rising crescendo of criticism, suggests that such features can be explained, not by a concept of urbanism, but by a 'culture of poverty' which is provincial, locally orientated and found both in the city and the rural area.[1] Indeed, his conclusion is that one of the priorities for urban research should be the delineation of distinctive sub-areas within the city in terms of demographic social, ecological and economic characteristics and the study of aspects of urbanization within such areas rather than within the city as a single unit. This indeed would appear to be the direction in which urban research should be developed.

The second conclusion from the present study relates to the role of the age structure in any theory of urban social structure. The importance of age composition was emphasized in the survey of attitudes to education within Sunderland. Parameters measuring aspects of age structure were also isolated as being of great importance in forming one of the four components derived from the multivariate analysis. The combination of large numbers of very young children and few women in the labour force together with fertility measures (which were compounded of aspects of the age structure) were grouped together as component 4 to form a composite measure of age composition which not only distinguished various council-estate areas, but also differentiated the two types of highly subdivided areas in the private housing sector of the town. Such emphasis on the importance and complexity of the age structure and life cycle is admirably supported by Sweetser's findings in Helsinki where no fewer than three of his six factors relate to different stages of the family life cycle, leading him to suggest that, 'the variables of aging and the passage of time ought to be more self-consciously considered in the theoretical deliberations of social ecologists'.[2]

From evidence elsewhere this important role which is played by the age structure can clearly be seen. Franklin, for example, has shown its diagnostic value in differentiating various communities within New Zealand.[3] At a micro scale, Jones has shown the age-structure variations between Catholic and Protestant populations in sub-areas of Belfast.[4] In such studies, age structure has been regarded as an important index of a range of social and economic characteristics. Gans, however, in arguing

---

[1] O. Lewis, 'Further observations on the folk-urban continuum and urbanization with reference to Mexico City', in Hauser and Schnore (eds.), *The study of urbanization*, pp. 491–503.
[2] F. L. Sweetser, 'Factorial ecology, Helsinki, 1960', *Demography*, 2 (1965), 384.
[3] S. H. Franklin, 'The age structure of New Zealand's North Island communities', *Econ. Geogr.* XXXIV (1958), 64–79.
[4] E. Jones, *A social geography of Belfast*, pp. 146–58.

against the concepts of Wirth and the 'micro-determinism' of Schnore and Duncan, attributes a more positive role to the age structure, since he argues that the position of the family in its life cycle can have strong determinative effects upon the area of residence which the family will select.[1] The tendency for a predominance of young families to be found in peripheral areas is well known and indeed is well illustrated by the data for Sunderland, where both private and council peripheral areas tend to have large proportions of young children. The population composition of inner areas within towns can equally be explained partly in terms of age structure. Typically in such inner areas there is a wide mixture of types of people. Gans suggests that there are five groups which he designates as cosmopolites, the unmarried and childless, the ethnic villagers, the deprived and the trapped. The reasons underlying their central location are threefold: choice, economic disability and age structure. The young unmarried element illustrates the functioning of the age-structure factor since typically they are found in inner areas only in the early stages of their life-cycle and move to more peripheral areas when they marry and raise families. Alonso, too, places similar weight on the importance of age structure and life-cycle stage in considering aspects of the swash and backwash of population movements between the inner and peripheral parts of towns.[2] The tendency which has been noted in certain American and British towns for a reversal of the usual outward movement of population to occur with centripetal movement in the form, usually, of high-status apart-ment developments represents, besides the fashionableness of certain areas such as parts of inner London, the effect of age structure in changing the demands of families. As a family grows older and children leave their parents, the parents' needs for large amounts of space correspondingly diminish. Whether a family chooses to live in the inner or the peripheral parts of a town can therefore be determined by its assessments of the importance of land inputs as against commuting inputs and the balance of this equation of family budgeting will change depending on which stage of its life-cycle the family is in as much as upon its income level. It is the same mechanism which has an effect in producing the concentration of older-age people in inner working class areas where, often, old people exchange their options on peripheral council houses in exchange for centrally located property no matter what state the property might be in.

The age structure of a given area therefore appears to be one of the vital determinants of residential location and, further, as the Sunderland survey demonstrated, has important repercussions on the nature of social relations

[1] H. J. Gans, 'Urbanism and suburbanism as ways of life: a re-evaluation of definitions', in A. Rose (ed.), *Human behaviour and social processes* (London, 1962), pp. 625–48.

[2] W. Alonso, 'The historical and the structural theories of urban form', *Land Economics*, XL (1964), 227–31.

within a given area. If one were to isolate the most important elements in any theory of residential structure, social class would obviously be one such element together with its many ramifications, but the second element would be the age structure, or the position of families within the family life-cycle. These two factors taken together appear to go a long way in explaining variations in the choice of residential area amongst urban populations. Given a choice of location, its social class and life-cycle stage define a set of role expectations which must influence a family in its choice of housing and of location within a town. In the Sunderland analysis, these two factors of class and age structure emerged from the statistical handling of the mass of data as components 1 and 4, and it would appear that there is both empirical and theoretical justification in considering them as basic to any understanding of the residential structure of urban areas.

A third conclusion might be underlined: namely the importance of the effects of the milieu on urban social structure. Within the term 'milieu' is included not only the physical environment of housing conditions, but also the less tangible factors of room density and area density, of the whole complex of such psychological factors as attitudes and of spatial location relative to facilities within the town and other people or types of people within a given local area. Such presumed influences have given rise to much controversy within the literature on social structure and social behaviour. The dispute arises not only from the general distaste for what is, often erroneously, considered to be untempered determinism, but also because of the difficulty of conducting controlled experiments which might effectively demonstrate the existence or absence of such influences. On the question of determinism, the inflexible postures which were adopted during the controversy over determinism/possibilism seem happily to have thawed with the realization that the conceptual gap between two apparently irreconcilable positions is no wider than the pin on which the philosopher's angels danced. In the present work a probabilistic model of behaviour has been adhered to by which the relationship between environment and behaviour, or environment and attitude, has been traced through the individual's perceptions and reactions to the 'real world' of eternal stimuli, bearing in mind the values and norms which are derived from his social class roles. From the results of the questionnaire survey, the most striking finding was the effect of the varying types of social composition of particular areas on the development of attitudes towards education. No matter what the area, the attitudes of individual families were more similar to those prevailing around them than to those of their 'objective' social class. The area of residence is therefore either a clue to or a determinant of these attitudes. The age structure and the degree of crowding, for example, were suggested as being of some importance as mechanisms by which attitudes could be formed or preserved. In the case of Lower Hendon, the importance of what the Sprouts call the 'psycho-milieu'[1] was stressed since, despite the relatively large number of children in the area, the reactions to this external fact differed markedly in Lower Hendon as compared to other working class areas. In such ways the import-ance of the milieu might be emphasized. Certainly the results of the attitude survey provided strong evidence that areal forces were operating.

On the other hand, it is interesting to note that little difference was found between the attitudes of working class populations living in the old impover-ished inner areas of the town and those in the new modern council estates. This is a conclusion which sits oddly with the emphasis which has been given to the role of the milieu, since one might have anticipated that improved housing would have led to the development of individualistic traits and so to greater aspiration. Partly, of course, the explanation might be found in the fact that room densities are just as high in council areas as in the older private areas, but also important is the fact that, while the external physical environ-ment of housing might have changed in some respects, many of the less tangible aspects of the milieu of the council estate remain unchanged.[2] The fact that many of those studies which have attempted to measure the changes which are involved in the move from inner redevelopment areas to new council housing areas have ignored the whole complex of the total milieu, perhaps accounts for their often blithe conclusions that real changes do occur. While it is difficult to control all the variables involved and so make a valid comparison of 'controlled' and experimental populations, rehousing often involves changes in important intervening variables such as class and the family cycle. The fact that such changes have too often been ignored makes one somewhat suspicious of certain of the conclusions that rehousing leads to the formation of new networks and new patterns of social behaviour and relationships. The evidence of longer-range studies, such as Willmott's Dagenham survey, appears to suggest that many of the alleged changes may be the product of short-term readjustment and that the basic conditions of life change very little as between the old and the new housing areas. Gold-thorpe and Lockwood come to a similar conclusion in their study of *embourgeoisement*.[3] The evidence of the Sunderland results would suggest that, as far as attitudes are concerned, the existence of better housing facilities alone makes little difference to parental approaches to education.

[1] H. and M. Sprout, *Ecological perspective on human affairs.*
[2] See, for example, A. Davis, 'Motivation of the underprivileged worker', in W. F. Whyte (ed.), *Industry and society* (New York, 1946), pp. 84–106.
[3] J. H. Goldthorpe, D. Lockwood *et al.*, 'The affluent worker and the thesis of *embourgeoisement*: some preliminary research findings', *Sociology*, 1 (1967), 11–31. Also Willmott, *The evolution of a community.*

Perhaps the conclusion which needs strongest emphasis is that studies of relocation need to consider all of the many complex elements which form the milieu before convincing conclusions can be made.

Such are three of the conclusions which might be drawn from the present study. At a more general level there is also an important methodological conclusion to be underlined. In the first chapter the links between geography and human ecology in their approach to the city were traced and their mutual overlaps with sociology were briefly examined. What has been attempted is a blending of the various approaches of a number of specialisms. In reviewing research work in urban geography, Berry concludes that we must not be sanguine about our knowledge of the residential structure of cities. Many sociologists and ecologists have recently made the same point. Berry, however, makes the further suggestion that, in view of the importance of ecological theories, such as those of Duncan or Shevky, the geographer must take second place to the sociologist in studying the residential patterning of cities.[1] In the present work, the importance of a sociological understanding has been stressed, but a more comprehensive approach has been adopted. The spatial interests of geography and the structural interests of sociology have been fused in an attempt to explore some of the relationships between space and social structure. What one hopes is important is the light which was shed on aspects of the structure of a British city, rather than any strict attribution of the treatment to one discipline or another. Preoccupations with the minutiae of '-ologies' and '-ographies' have too often robbed one discipline of the benefit of the ideas and findings of another to the mutual disadvantage of both.

In trying to work towards an understanding of the multiple strands which make up urban existence, scholars in each of these various disciplines inevitably overlap in their interests even though their detailed focus may differ. The geographer's focus in studying the city must always be spatial and there is a great deal of fascinating and valuable material from the work of sociologists which can help in our understanding of the spatial element in urban social structure, and which is thus relevant to a geographical interest in the urban area. In particular, two aspects of location within the urban area have been studied by sociologists. First is the role of various determinants of intra-city location, a topic which has been at the core of geographical concern and to which research from other fields could be used more fruitfully by geographers. An excellent start to the formulation of a scheme of residential location has, for example, been made by Duncan in her study of the relationships of work potential and residence patterns in Chicago.[2] The second aspect

[1] Berry, *op. cit.* in Hauser and Schnore (eds.), *The study of urbanization.*
[2] B. Duncan, 'Variables in urban morphology', in Burgess and Bogue (eds.), *Contributions to urban sociology*, pp. 17–30.

—the role of location within social structure—is one which deserves much greater attention in urban studies and which has guided part of the present work. From the available field, two topics will be singled out to demonstrate their spatial focus: the evidence of marriage patterns, and the evidence of location itself as a determinant of social networks.

The studies of marriage patterns have shown remarkably consistent results. From Bossard's original work in Philadelphia, in which he showed that the proportion of marriages declined steadily as the distance between the two contracting parties increased, the subject was taken up by Davie and Reeves in New Haven and, by 1958, Katz and Hill were able to collect for review no fewer that fourteen studies relating to propinquity and marriage.[1] All of them demonstrate that the percentage of marriages falls as one moves from people living in the same block to those living further apart. On the basis of the evidence, Katz and Hill link together the three factors of residential segregation, marital endogamy (with respect to status, ethnic type and so forth) and spatial propinquity in marriage partners and suggest that the interaction of the three is a mechanism by which social and residential cleavages in an urban society can be perpetuated. In a more rigorous study than any of the earlier ones, Ramsøy, working on Oslo material, has recently carried this examination of the interrelationship of the three factors a stage further by looking at the expected and actual patterns of marriage rates holding constant each of the three factors in turn.[2] While she shows that the percentages of marriages decline regularly with increasing social distance and with increasing physical distance, each of these effects operates independently; in other words spatial propinquity has effects in its own right and not simply because of the residential segregation of the city.

Such a conclusion ties in very closely with the second aspect of spatial location—its effect upon the formation of social networks. A number of studies have shown that people living close to one another in terms of distance, accessibility or physical orientation tend to become friends or form close-knit social groups[3] and, while this conclusion has to be viewed in the light of the homogeneity of the areas studied,[4] their findings complement the studies of marital propinquity in an interesting fashion. The most remarkable

[1] J. H. S. Bossard, 'Residential propinquity as a factor in marriage selection', *Am. J. Sociol.* xxxviii (1932), 219–24; M. R. Davie and R. J. Reeves, 'Propinquity of residence before marriage', *Am. J. Sociol.* xliv (1939), 510–17; A. M. Katz and R. Hill, 'Residential propinquity and marital selection: a review of theory, method and fact', *Marriage and family living*, xx (1958), 27–35.

[2] N. R. Ramsøy, 'Assortive mating and the structure of cities', *Am. Sociol. Rev.* xxxi (1966), 773–86.

[3] T. Caplow and R. Forman, 'Neighbourhood interaction in a homogeneous community', *Am. Sociol. Rev.* xv (1950), 357–67; Form, 'Stratification in...housing areas', *J. Social Issues*, vii (1951); Whyte, *The organization man*; Festinger, Schachter and Back, *Social pressures in informal groups* (New York, 1950).

[4] Gans, 'Planning and social life', *J. Am. Inst. of Planners*, xxvii (1961).

of these studies is the work of Festinger, Schachter and Back on a new housing project for students at the Massachusetts Institute of Technology in which they study the social-psychological consequences of residential proximity. Their sociometric tests demonstrate that, even at a micro scale, distance is an important variable in determining who is friendly with whom. Within the blocks of part of the settlement, acquaintanceships were influenced by the distance between doorways and the positioning of staircases; within the houses of another part of the settlement, they were determined by the functional distances between houses and the placement of the houses themselves within each self-contained 'court'. The social networks thus formed by the physical layout of residences led to polarization of attitudes around different groups of houses and an experiment in rumour planting illustrated that channels of communication are related to this sociometric structure and particularly to the cohesiveness of a court as revealed by the sociometric tests.

It was largely upon the basis of such research on marriage patterns and the effects of distance that Beshers developed his argument of the relations of space and social structure.[1] In attempting to explore the nature of the relationship and so blend the approaches of ecology and functional sociology, he explains the distribution of social characteristics within towns in terms of two factors which he calls 'prestige attraction' and 'social distance', which are the positive and negative aspects of a more general principle, 'social desirability'. In this there are at least three mechanisms at work: differential income, preference for similar neighbours, and rejection of dissimilar neighbours. The first of these is well documented; the latter two are illustrated by the willingness of white collar workers to pay more for prestige locations and the fact that negroes are often forced to pay excessive amounts for any location at all. Given these broad principles of spatial distribution, he then deduces their consequences in terms of probabilities of informal contact similar to Festinger's thesis. In the heterogeneous community both social and ecological factors have to be considered in building up a theory of group formation. Eligible mates are culturally defined; the probability of marriage is related to the probability of contact; the probability of contact is limited by the spatial distribution of opportunities or eligible marriage partners. Thus he concludes, 'If the main lines of this analysis are correct, we may think of a stratification system as a self-maintaining system. We know that the location of social characteristics can be predicted in terms of a stratification theory, especially in terms of social distance, and we also know that location itself is a determinant of the subsequent social relationships that will make up the future stratification system.'[2]

[1] Beshers, *Urban Social Structure.* [2] *Ibid.* p. 125.

Given the geographer's prime interest in location and spatial processes, the evidence of such sociological work is obviously of great relevance. It is in the blending of the spatial and non-spatial elements within urban social structure, which is so well illustrated by Besher's approach, that holistic explanations of residential location and its connections with social structure can best be developed. It is in the mutual interests of the various disciplines which impinge on urban studies that their ideas and concepts should be pooled and be more widely familiar to a variety of scholars in related fields. Now that geographers are becoming less inhibited in their use of social data and now that population geography has become an accepted part of the discipline, there is great scope for further fusion of the concepts of geography and sociology. Armed now with a battery of techniques and concepts, the geographer should find the urban area pregnant with research potential. One of the prerequisites of such geographical analysis of urban society is sound comparative study of a wide cross-section of British urban areas. We still know remarkably little about the patterns of distribution and the patterns of association of phenomena within towns. The multivariate studies which have been reviewed here suffer from the fact that the variables used in the analyses were not identical. Before we can confidently isolate important aspects of urban social structure, these basic comparative distributions must clearly be enumerated.[1] For such a project, multivariate techniques would appear to be admirably suited and increasing availability of and access to census material at finer areal scales would be axiomatic.

Given such preliminary understanding of the facts of urban phenomena, the research topics which come to mind for geographical study are those for which a spatial viewpoint should provide new sources of theory. One which is suggested by the present work is the analysis of differential fertility within the urban area. A framework of urban social areas would be an ideal design within which to test hypotheses relating to fertility differences within new estates as against inner urban areas or to the changing pattern of fertility differentials which is asserting itself within the social classes throughout the country. Likewise, an areal framework would be well suited to the geographical study of marriage patterns, a field which has been tackled in America, but in which British scholars have shown little interest.[2] Yet marriage patterns

[1] The Inter-University Census Tract Committee (see note above, p. 40), is, in fact, in process of conducting such a comparative urban analysis. Such work will complement the growing number of multivariate studies of towns such as Seattle, Newark, Toronto, Boston, Helsinki, Copenhagen, Cairo and Canberra in addition to the British studies.

[2] A recent geographical study, which illustrates the ever-increasing social scope of the subject, is R. Morrill and F. R. Pitts, 'Marriage, migration and the mean information field: a study in uniqueness and generality', *Ann. Ass. Am. Geogr.* LVII (1967), 401–22. In studying data for individuals and linking the spatial pattern which are found to a general theory of information fields, this is an excellent example of the direction in which one hopes that geographical study of social data might develop.

and the choice of marriage partner are basic underpinnings not only of social structure but also of aspects of internal population movements within an area. A deeper understanding of their functional mechanisms would clarify many aspects of urban systems. Thirdly, the study of life-cycle variations as a factor in the selection of urban residential location is a topic needing more intensive study and one which is likely to suggest important conclusions in relation to the residential structure of cities. Again, a blending of sociological and geographical expertise would be invaluable in testing the extent to which age structure plays a part in determining choice of residence.

The relationship between spatial and social processes is complex, but of great importance because, in trying to tease out some of its strands, we shall learn a great deal about the composition of urban society. A vital aspect of location in the study of social structure is that it helps to determine the probability of contact between individuals. If we can discover the consequences of the probability of contact upon social relationships, then we can establish a link between environment and social structure. The spatial viewpoint of geography offers a fruitful methodology through which such clearer understanding might be achieved.

APPENDICES

BIBLIOGRAPHY

INDEX

VALUATION BOOK MATERIAL

Three aspects will be discussed: the terminology of valuation, the use of the measure of subdivided housing and the use of median values.

TERMINOLOGY OF THE VALUATION LISTS

From the data found in the Valuation Books, the gross value of property is the parameter used for each of the three dates. Where reference is made in the text to 'rating values' or 'valuations', it is this gross value which is referred to. It is essential at the outset to clarify this term in relation to others used in connection with the valuation of property.

'*Gross value*' is essentially a measure of hypothetical rental which is calculated for all property whether an actual rent is paid or not. This interpretation has applied from the very early days of rating procedure since, although the first explicit reference to annual rent as a measure of value appeared only in the Parochial Assessments Act of 1836, the object of this Act does not seem to have been to introduce any new principle, but to ratify already established practice. The data which are used for Sunderland from the 1892 Valuation Books are the gross estimated rentals, which were defined by Section 15 of the Union Assessment Committee Act of 1862 as 'the rent at which the hereditament might reasonably be expected to let from year to year, free of all the usual tenant's rates and taxes and tithe commutation rent charges, if any'. Gross value in more recent times is little different. As defined by Section 68 of the 1925 Rating and Valuation Act, it is 'the rent at which a hereditament might reasonably be expected to let from year to year if the tenant undertook to pay all the usual tenant's rates and taxes . . . and if the landlord undertook to bear the cost of the repairs and insurance, and other expenses, if any, necessary to maintain the hereditament in a state to command that rent'. Thus, while the allocation of detailed costs as between tenant and landlord has changed over time, the gross value has always been an estimate of hypothetical annual rental.

'*Rateable value*' is today, in almost all cases, synonymous with the net annual value of property. It is the figure from which the local rates are calculated by multiplying the rateable value by the poundage of the rate made by the local authority. This rateable value is either calculated independently of the gross value or, more usually, is calculated by a specified reduction of the gross value. In the 1963 revaluations, the rateable value was calculated by using a variable fraction whose size depended upon whether the gross value of the property was under £55 (in which case the rateable value becomes 45 per cent of the gross value), was over £55 but less than £430 (in which case the rateable value is £25 plus one-fifth of the amount by which the gross value exceeds £55), or was over £430 (in which case the rateable value becomes £100 plus one-sixth of the amount by which the gross value exceeds £430). It was partly because a variable fraction is used to calculate the rateable value that the gross value was here used in preference.

Confusion must be avoided between the gross value of a property and the actual rent which might be paid for it. The gross value is a purely hypothetical measure of rent. The principles of valuation include the injunction that 'the actual rent at which a hereditament is let or the actual rent at which similar hereditaments in similar economic sites are let so that they are truly comparable, are not necessarily conclusive evidence, but may be the best evidence, of value'.[1] Ryde quotes Lord Atkinson: 'This imaginary rent is not to be confounded with the rent which an actual tenant in possession in fact pays.' Indeed, in practice it would appear that there are discrepancies between the gross value of property and the actual rents charged for property, as the Milner Holland Report has shown.[2] However, this in no way invalidates the use of gross rating values as an index of social class. In the present confused state of the housing market, which is complicated by the existence, cheek by jowl, of controlled and uncontrolled properties, the hypothetical rental which is measured by the gross rating value would seem to be a more reliable index than actual rent.

SUBDIVISION OF HOUSES

In describing properties, the valuation lists often draw a distinction between 'houses' on the one hand, and 'flats' and 'rooms' on the other. It is this distinction upon which the degree of subdivision of housing has been based: 'houses' have been taken to be unsubdivided, one-family dwelling places. However, reservations as to the accuracy and consistency of the valuation definitions should be recorded. The criteria of the definitions are nowhere made explicit, but in practice a household space is recorded separately in the valuation records only if it is a self-contained entity. If a house has been divided into a number of sub-units each of which has its own cooking and washing facilities, each sub-unit will probably be recorded separately. One adds 'probably' advisedly since there appear to be few set rules governing these descriptions and one has to accept that some separate households are not revealed by the valuation list descriptions. The existence of such 'hidden' households (which presumably have lower income levels than would be suggested by the gross value of the whole property) can be inferred from the Register of Electors. In Sunderland, for example, within such areas as the rooming-house district of Bishopwearmouth, the Register of Electors shows that some of the so-called 'houses' of the valuation lists are in fact lived in by a number of adults with

[1] *Ryde on rating : the law and the practice* (11th edition, London, 1963), p. 371.
[2] *Report of the committee on housing in Greater London* (Cmd. 2605, H.M.S.O. London, 1965), pp. 346–57.

different surnames, and thus presumably in some of such cases by a number of separate households.

In some respects, census data are more satisfactory than the valuation list material since the census does attempt to apply a standard definition of a 'dwelling' ('a building or part of a building which provides separate living quarters'). Where houses have been adapted to accommodate more than one family, the census definition would record the sub-units as 'dwellings' only if they were structurally separate (i.e. with a separate access to the street and a solid partition between it and an adjacent dwelling). Where access to the unit could be made only through part of the building, it would be counted as a dwelling if such access was by means of a common landing or staircase and movement between its rooms would have to be possible without using common landings or spaces.[1] These census definitions would therefore exclude many of the rooming-house subdivisions, but would include those subdivisions in which structural alterations to once self-contained buildings met the requirements set out above.

The census data can therefore provide no more accurate a measure of the historical process of subdivision than that provided by the valuation lists and, in fact, the distribution of dwellings with more than one household space, as given by census e.d. data for Sunderland, showed a pattern similar to, but less extensive than, the distribution of 'rooms' given by the valuation lists. The valuation list material was therefore taken as being a more discriminating guide to the historical process of subdivision.

USE OF MEDIAN VALUES

In assigning median values to each street or part of a street, a certain amount of internal heterogeneity has inevitably been concealed. By making the sub-areas delineated on the basis of rating values as small as possible, this problem was avoided in all but the rooming-house areas in Bishopwearmouth where, within individual streets, there are often large differences in value between those houses which have been subdivided and in which individual household valuations are low, and those which have not been subdivided and in which individual valuations may be very high. Any measure of central tendency would conceal such variations. Thus, for example, a street which comprises 10 undivided 'houses' each valued at £150 and 6 houses subdivided into 12 'flats' or 'rooms' each valued at £40, will have a median value of £40. A second street composed of 11 'houses' each valued at £150 and 5 houses divided into 10 'flats' each valued at £40, will have a median value of £150. However, the use of averages in place of medians would not solve the difficulty since, as an estimating figure, average values would be representative of neither type of housing. Ideally, the measure of central tendency should be supplemented by some measure of dispersion, but this could neither be conveyed graphically nor could it be used in the statistical handling of the data. Fortunately, the areas in which such problems were apparent were few, being restricted to a number of streets in the Bishopwearmouth area.

[1] *Census, 1961*, General Explanatory Notes, pp. x–xi.

APPENDIX B

Product-moment correlation matrices for (a) the private housing and (b) the whole town

Variable	1	2	3	4	5	6	7	8	9	10
1 Percentage social class I	—	0·435	-0·612	-0·596	0·881	-0·826	-0·024	-0·399	0·655	-0·255
2 Percentage social class II	0·494	—	0·551	-0·601	0·712	-0·641	0·040	-0·375	0·568	-0·416
3 Percentage social class III	-0·592	-0·550	—	0·027	-0·526	0·640	-0·000	0·327	-0·410	0·153
4 Percentage social class IV	-0·579	-0·593	-0·086	—	-0·839	0·613	0·006	0·357	-0·606	-0·435
5 'Social Class Score'	0·883	0·749	-0·459	-0·834	—	-0·865	-0·029	-0·472	0·728	-0·428
6 Percentage terminal education age below sixteen	-0·825	-0·689	0·594	0·600	-0·863	—	0·104	0·468	-0·645	0·367
7 Percentage females working	-0·087	0·050	0·033	0·079	-0·104	0·068	—	-0·088	0·018	0·129
8 Percentage females ever married	-0·047	0·021	0·029	0·041	-0·055	0·097	-0·300	—	-0·239	0·551
9 J-index	0·369	0·252	-0·132	-0·349	0·390	-0·337	0·300	-0·271	—	0·166
10 Percentage aged 0-14	-0·362	-0·482	0·289	0·381	-0·478	0·445	-0·040	-0·035	0·322	—
11 Percentage aged 20-24	-0·094	-0·083	-0·033	0·171	-0·139	0·016	0·050	0·300	-0·400	-0·291
12 Percentage aged 65 and over	0·292	0·406	-0·240	-0·322	0·404	-0·372	-0·192	0·147	-0·458	-0·846
13 Fertility ratio 1 (children 0-4)	-0·121	-0·195	0·092	0·155	-0·176	0·106	-0·196	0·466	-0·383	0·041
14 Fertility ratio 2 (children 0-9)	-0·343	-0·463	0·243	0·393	-0·462	0·424	-0·274	0·307	-0·040	0·773
15 Percentage owner-occupiers	0·670	0·639	-0·383	-0·642	0·775	-0·677	-0·116	0·151	0·235	-0·642
16 Percentage council	-0·394	-0·393	0·277	0·340	-0·450	0·389	0·211	-0·507	0·285	0·635
17 Percentage renting furnished	0·138	0·171	0·184	-0·700	0·157	-0·266	-0·138	0·318	-0·264	-0·241
18 Percentage renting unfurnished	-0·174	-0·139	0·043	0·203	-0·203	0·209	-0·158	0·574	-0·652	-0·205
19 Percentage one-person households	0·061	0·150	-0·136	-0·042	0·102	-0·137	-0·124	0·393	-0·562	-0·551
20 Percentage two-person households	0·406	0·454	-0·292	-0·386	0·491	-0·435	-0·136	0·345	-0·308	-0·813
21 Percentage six-person households	-0·388	-0·457	0·247	0·413	-0·488	0·392	0·248	-0·446	0·160	0·593
22 Average no. of persons per room	-0·624	-0·671	0·390	0·630	-0·759	0·684	0·031	-0·108	-0·165	0·736
23 Percentage households at over 1·5 persons per room	-0·446	-0·514	0·188	0·565	-0·589	0·481	-0·089	0·179	-0·460	0·403
24 Median gross rateable value	0·661	0·496	-0·404	-0·511	0·684	-0·712	0·095	-0·423	0·730	-0·120
25 Percentage dwellings with over eight rooms	0·304	0·173	-0·275	-0·121	0·271	-0·382	-0·130	0·115	-0·111	-0·187
26 Percentage unsubdivided dwellings	0·192	0·198	0·005	-0·308	0·267	-0·216	0·167	-0·528	0·495	-0·041
27 Percentage dwellings with one household space	0·051	0·041	0·095	0·165	0·096	-0·081	0·132	-0·397	0·293	0·005
28 Percentage households sharing dwellings	-0·027	-0·027	-0·090	0·128	-0·064	0·053	-0·149	0·443	-0·326	-0·053
29 Percentage without fixed bath	-0·238	-0·246	0·130	0·250	-0·287	-0·328	-0·207	0·583	-0·647	0·123
30 Percentage with all four amenities	0·181	0·152	-0·091	0·170	0·199	-0·224	0·234	-0·616	0·663	0·218

private matrix (30 variables × 159 e.d.s.)

total matrix (30 variables × 263 e.d.s.)

NOTE. Upper segment refers to the private housing sector only (159 enumeration districts)

Lower segment refers to the whole town, including council areas (263 enumeration districts).

Variable	11	12	13	14	15	16	17	18	19	20
1 Percentage social class I	-0·160	0·061	-0·310	-0·387	0·630	-0·109	-0·023	-0·636	-0·324	0·312
2 Percentage social class II	-0·195	0·245	-0·390	-0·521	0·589	0·003	0·014	-0·603	-0·205	0·361
3 Percentage social class III	0·077	-0·032	0·266	0·273	-0·333	-0·001	-0·082	0·362	0·156	-0·142
4 Percentage social class IV	0·226	-0·266	0·334	0·501	-0·691	0·073	0·080	0·684	0·261	-0·420
5 'Social Class Score'	-0·224	0·221	-0·402	-0·546	0·755	-0·088	-0·025	-0·761	-0·313	0·427
6 Percentage terminal education age below sixteen	0·197	-0·181	0·270	0·485	-0·647	0·174	-0·138	0·668	0·192	-0·350
7 Percentage females working	-0·078	-0·081	-0·003	0·182	0·055	0·061	-0·089	-0·034	0·044	0·117
8 Percentage females ever married	0·178	0·472	0·267	0·562	-0·421	-0·073	0·156	0·436	0·009	-0·252
9 J-index	-0·203	-0·171	-0·453	-0·400	0·818	-0·050	-0·227	-0·800	-0·571	0·291
10 Percentage aged 0-14	0·325	-0·736	0·304	0·873	-0·501	0·076	0·020	0·504	-0·154	-0·687
11 Percentage aged 20-24	—	-0·153	0·150	0·378	-0·350	-0·032	0·119	0·375	0·155	-0·250
12 Percentage aged 65 and over	0·166	—	-0·186	-0·465	0·190	-0·074	0·049	-0·187	0·432	0·425
13 Fertility ratio 1 (children 0-4)	0·299	0·094	—	0·353	-0·483	-0·123	0·327	0·426	0·369	-0·197
14 Fertility ratio 2 (children 0-9)	-0·220	-0·455	0·291	—	0·587	0·039	0·014	0·598	0·092	0·598
15 Percentage owner-occupiers	-0·064	-0·554	-0·056	-0·485	—	-0·096	-0·341	-0·944	-0·453	0·564
16 Percentage council	-0·297	-0·667	-0·342	0·298	-0·719	—	-0·159	-0·025	-0·112	-0·150
17 Percentage renting furnished	0·538	0·290	0·396	-0·106	0·078	0·409	—	0·221	0·389	-0·054
18 Percentage renting unfurnished	0·476	0·349	0·509	0·106	-0·027	-0·653	0·408	—	0·447	-0·569
19 Percentage one-person households	0·353	0·711	0·405	-0·175	0·229	-0·614	0·478	0·623	—	-0·043
20 Percentage two-person households	0·180	0·789	0·168	-0·445	0·711	-0·747	0·274	0·289	0·588	—
21 Percentage six-person households	0·001	-0·613	0·185	0·282	-0·662	0·727	-0·254	-0·326	-0·541	-0·776
22 Average no. of persons per room	0·005	-0·648	0·129	0·602	-0·807	0·541	-0·129	0·089	-0·323	-0·682
23 Percentage households at over 1·5 persons per room	0·244	-0·262	0·348	0·495	-0·549	0·084	0·139	0·452	0·054	-0·330
24 Median gross rateable value	-0·294	-0·023	-0·398	-0·353	0·475	0·108	-0·154	-0·673	-0·353	0·029
25 Percentage dwellings with over eight rooms	0·388	0·257	0·266	-0·125	0·118	0·292	0·612	0·236	0·303	0·188
26 Percentage unsubdivided dwellings	-0·417	-0·165	-0·439	0·299	0·250	0·311	-0·446	-0·714	-0·586	0·106
27 Percentage dwellings with one household space	-0·312	-0·121	-0·248	0·172	0·068	0·316	-0·326	-0·548	-0·353	-0·094
28 Percentage households sharing dwellings	0·381	0·178	0·300	0·145	-0·046	-0·372	0·438	0·593	0·422	0·144
29 Percentage without fixed bath	0·340	0·283	0·507	0·211	-0·034	-0·563	0·262	0·838	0·547	0·280
30 Percentage with all four amenities	-0·439	-0·390	-0·551	-0·135	-0·026	0·641	-0·460	-0·872	-0·648	-0·354

Product-moment correlation matrices for (a) the private housing and (b) the whole town (cont.)

Variable	21	22	23	24	25	26	27	28	29	30
1 Percentage social class I	-0·321	-0·581	-0·509	0·803	0·206	0·418	0·224	-0·234	-0·621	0·631
2 Percentage social class II	-0·377	-0·652	-0·629	0·638	0·076	0·425	0·197	-0·212	-0·647	0·600
3 Percentage social class III	0·145	0·362	0·260	-0·564	-0·237	-0·194	0·006	0·023	0·433	-0·435
4 Percentage social class IV	0·432	0·692	0·705	-0·667	-0·026	-0·514	-0·370	0·357	0·640	-0·603
5 'Social Class Score'	-0·433	-0·754	-0·703	0·867	0·168	0·530	0·305	-0·310	-0·754	0·732
6 Percentage terminal education age below sixteen	0·349	0·671	0·568	-0·854	-0·322	-0·431	-0·236	0·235	0·718	-0·668
7 Percentage females working	-0·086	-0·088	-0·116	0·023	-0·101	0·105	0·071	-0·071	-0·116	0·144
8 Percentage females ever married	0·082	0·507	0·466	-0·524	-0·071	-0·468	-0·351	0·380	0·516	-0·536
9 J-index	-0·397	-0·583	-0·611	0·807	-0·078	0·587	0·294	-0·327	-0·758	0·803
10 Percentage aged 0-14	0·518	0·729	0·721	-0·400	-0·018	-0·513	-0·370	0·352	0·510	-0·426
11 Percentage aged 20-24	0·250	0·356	0·444	-0·242	0·063	-0·373	-0·334	0·358	0·340	-0·331
12 Percentage aged 65 and over	-0·284	-0·478	-0·384	0·161	0·124	0·189	0·153	-0·124	-0·177	0·073
13 Fertility ratio 1 (children 0-4)	0·308	0·411	0·480	0·399	0·206	-0·378	-0·157	0·203	0·432	-0·492
14 Fertility ratio 2 (children 0-9)	0·456	0·764	0·791	-0·556	-0·067	-0·572	-0·428	0·413	0·691	-0·618
15 Percentage owner-occupiers	-0·618	-0·789	-0·827	0·818	-0·130	0·797	0·449	-0·484	-0·760	0·799
16 Percentage council	0·203	0·073	0·004	-0·072	-0·154	0·011	0·058	-0·078	-0·041	0·080
17 Percentage renting furnished	0·126	0·139	0·238	-0·129	0·568	-0·431	-0·234	0·350	0·057	-0·304
18 Percentage renting unfurnished	0·589	0·763	0·793	-0·819	0·074	-0·797	-0·482	0·504	0·760	-0·785
19 Percentage one-person households	0·051	0·144	0·269	-0·401	0·186	-0·504	-0·249	0·298	0·394	-0·519
20 Percentage two-person households	-0·635	-0·565	-0·579	0·367	-0·057	0·448	0·322	-0·310	-0·357	0·332
21 Percentage six-person households	—	0·560	0·599	-0·336	0·147	-0·372	-0·233	0·252	0·241	-0·238
22 Average no. persons per room	0·612	—	0·919	-0·750	-0·114	-0·623	-0·372	0·384	0·736	-0·683
23 Percentage households at over 1·5 persons per room	0·344	0·790	—	-0·697	0·060	-0·705	-0·422	0·443	0·753	-0·716
24 Median gross rateable value	0·008	-0·504	0·595	—	0·157	0·681	0·406	-0·425	-0·865	0·863
25 Percentage dwellings with over eight rooms	-0·168	-0·250	0·006	0·116	—	0·254	-0·391	0·445	0·077	-0·071
26 Percentage unsubdivided dwellings	0·166	-0·227	-0·453	0·605	-0·289	—	0·621	-0·656	-0·668	0·750
27 Percentage dwellings with one household space	0·153	-0·091	-0·301	0·406	-0·450	0·579	—	0·779	0·453	0·478
28 Percentage households sharing dwellings	-0·190	0·061	0·303	-0·422	0·507	-0·617	-0·976	—	0·460	-0·520
29 Percentage without fixed bath	-0·343	0·166	0·479	-0·764	0·097	-0·630	-0·524	0·552	—	-0·938
30 Percentage with all four amenities	0·395	-0·063	-0·417	0·725	-0·231	0·687	0·548	-0·606	-0·955	—

CALCULATION OF COMPONENT SCORES

The following short-cut procedure for calculating the component scores for each enumeration district was kindly suggested by Miss Elizabeth Gittus (Newcastle University).

For each enumeration district, the original raw data have to be standardized and multiplied by the weights of the latent vectors of each variable for each component in turn. This can be expressed as follows:

$$\sum_{i=1}^{N} l_i \frac{x_i - \bar{x}_i}{\sigma_i}$$

where $\bar{x}_i$, σ_i are the mean and standard deviation of the ith variable and l is the latent vector. Where the elements appropriate to the latent vectors are calculated so that $\Sigma\, l_i^2 = \lambda$, to achieve unit variance we have to calculate the following:

$$\sum_{i=1}^{N} \frac{l_i}{\lambda} \frac{x_i - \bar{x}_i}{\sigma_i}$$

This can be broken down into

$$\sum_{i=1}^{N} \frac{1}{\lambda} \left(\frac{l_i}{\sigma_i}\right) x_i \; - \; \sum_{i=1}^{N} \frac{l_i}{\lambda} \frac{\bar{x}_i}{\sigma_i}$$

The second half of this expression is a constant which can be evaluated for each component and which will be called C. Thus, for each variable i, the following factors have to be calculated for each component:

$$\frac{l_i}{\lambda} \frac{1}{\sigma_i} \quad \text{and} \quad \frac{l_i}{\lambda} \frac{1}{\sigma_i} \bar{x}_i$$

For the ith variable, these can be designated as A_i and B_i. Again $\Sigma B_i = C$. For each enumeration district, the component score is thus arrived at by calculating $(\Sigma A_i x_i) - C$ for each component, where x_i is the raw value of the ith variable.

Where the vector elements are initially calculated so that $\Sigma l_i^2 = 1 \cdot 0$, $\sqrt{\lambda}$ is substituted for λ in the above formulae.

This short cut saves a great deal of additional calculation by avoiding the need to standardize the original raw data.

APPENDIX D

COMPONENT SCORES FOR ENUMERATION DISTRICTS

A] PRIVATE HOUSING AREAS

Ward and enumeration district number	Component			
	1	2	3	4
Bishopwearmouth				
29	1·78	2·20	− 3·96	3·66
30	1·68	1·15	− 0·70	1·45
31	1·13	1·44	− 1·79	1·20
32	0·59	1·72	− 1·11	2·12
35	0·17	1·58	− 0·45	− 0·85
36	0·11	1·64	− 0·63	− 0·14
37	1·24	1·20	− 0·85	1·51
38	1·89	0·50	− 1·51	2·06
39	1·63	0·40	− 0·12	− 0·06
40	1·52	0·34	− 1·07	0·15
41	2·43	0·34	− 1·93	0·47
Bridge				
32	− 0·04	1·15	1·16	− 0·58
33	− 0·07	1·27	1·38	− 0·36
34	0·51	0·60	1·49	− 0·55
35	0·16	0·84	1·24	− 0·19
36	0·34	0·93	1·50	− 1·08
37	0·61	0·85	1·12	0·19
38	0·57	0·98	1·93	− 0·10
39	0·31	0·65	1·69	− 0·08
40	0·46	1·16	0·71	− 0·71
42	0·29	1·47	1·81	0·06
43	− 0·00	1·33	0·63	− 0·05
Central				
18	0·30	1·08	0·71	0·58
25	− 0·41	0·64	0·68	1·05
27	0·25	1·29	1·50	− 1·28
29	− 0·50	0·27	0·35	1·31
Colliery				
52	1·64	− 0·38	0·57	− 0·36
53	2·25	− 0·38	0·75	1·55
56	0·42	0·76	1·34	0·05
57	− 0·34	1·65	0·42	− 0·54
58	− 0·28	1·20	0·86	− 0·72
59	0·15	0·79	1·66	− 1·03

260

Ward and enumeration district number	Component			
	1	2	3	4

Ward and enumeration district number	1	2	3	4
Colliery (*cont.*)				
60	−0·18	0·81	1·14	0·55
61	0·56	0·85	0·99	−0·31
Deptford				
44	0·60	0·58	1·64	−0·91
45	0·38	0·80	1·31	−0·59
46	0·58	0·70	2·00	−0·47
47	0·38	0·72	1·64	−0·93
48	−0·24	1·31	0·73	−0·05
49	−0·16	1·28	0·97	−0·22
50	−0·17	1·19	1·06	−0·05
51	0·61	0·54	1·42	−0·50
52	0·31	1·08	1·23	−0·31
53	0·74	0·37	1·29	−0·32
54	0·68	0·30	1·40	−0·96
55	0·88	0·51	0·25	−0·33
56	0·79	0·72	1·72	−0·86
57	1·02	0·74	1·23	−0·86
Fulwell				
36	1·43	0·43	0·91	−1·12
37	1·72	−0·05	−0·13	0·37
38	1·81	−0·64	−0·06	−0·73
39	1·30	−0·49	0·72	−0·67
40	1·13	0·27	1·64	0·72
41	2·10	−0·75	−0·04	−0·87
42	1·61	−0·18	1·07	0·41
43	1·61	−0·35	0·65	−0·70
44	1·44	−0·56	0·97	0·07
45	2·38	−0·90	−1·14	−0·20
46	1·42	−0·69	−0·00	−1·15
48	1·06	−0·48	−0·13	−0·49
49	1·03	−0·99	−0·63	−1·95
50	1·19	−0·77	−0·15	−1·41
51	1·75	−0·71	0·51	−0·29
Hendon				
1	1·44	−0·65	0·51	−0·59
2	−0·14	1·01	0·51	0·04
3	0·15	0·46	1·44	0·85
6	2·04	−0·70	0·23	−0·16
7	1·40	0·59	−1·31	0·22
8	0·86	−0·22	0·75	0·17
9	1·00	−0·08	1·40	−0·24
10	0·67	0·23	1·62	−0·12
11	1·36	0·43	0·43	0·68
12	0·48	0·59	1·32	−0·36
Humbledon				
66	1·31	−0·88	−0·10	−0·04
67	2·44	−0·90	−0·69	−0·26
68	2·01	−0·78	0·12	−1·00
69	1·33	−1·10	−1·08	−1·97
Monkwearmouth				
2	0·61	0·53	1·96	0·73
3	0·44	0·72	1·71	−0·42

Ward and enumeration	Component			
district number	1	2	3	4

Monkwearmouth (*cont.*)

4	0·29	0·73	1·11	−0·18
5	0·42	0·48	1·09	−0·42
6	1·37	−0·03	1·08	−0·11
7	−0·64	1·10	0·06	−0·83
8	−0·79	1·20	−0·06	−0·14
11	−0·80	2·00	−1·00	−1·47
12	−0·73	1·42	0·10	−1·42
14	−0·41	1·47	−0·04	−0·91
15	−0·67	2·18	−1·65	−1·59
16	−0·28	2·21	−1·08	0·54
17	0·13	1·40	−0·36	0·01
18	0·42	1·25	1·27	−0·24
19	1·35	0·11	2·07	0·83
20	0·42	1·37	0·18	0·15
21	−0·40	1·91	−1·36	−0·00

Pallion

35	0·68	0·55	1·85	−0·28
40	0·81	0·31	1·68	0·74
41	1·20	−0·09	1·34	−0·31
42	2·24	−0·86	−0·19	0·40
43	0·90	−0·65	0·41	−0·22
44	−0·01	−0·10	1·25	1·58
45	−0·14	−0·09	1·05	1·20

Park

1	0·27	2·03	−1·23	1·00
2	−0·17	1·19	0·48	0·13
4	−0·45	2·51	−2·22	−0·09
6	−0·05	1·93	−1·51	0·09
7	0·03	1·07	0·62	−0·47
8	−0·02	2·04	−1·50	0·19
9	−0·32	1·38	1·00	−1·47
10	−0·25	1·94	−1·69	0·72
11	−0·36	1·31	0·55	−0·24
12	0·39	0·92	0·15	0·52
13	0·68	0·96	0·83	−0·70
14	−0·27	2·47	−1·83	−0·38
15	0·29	1·28	0·65	−0·34
16	−0·88	2·48	−3·40	0·06
17	−0·37	2·06	−1·25	−0·34

Roker

22	1·08	0·44	1·73	0·31
23	0·76	0·75	1·63	−0·55
24	0·68	0·59	0·97	0·05
25	0·56	1·24	0·58	0·37
26	0·69	1·18	−0·11	−0·24
27	1·41	0·55	−0·57	0·52
28	1·78	0·64	−1·21	2·23
29	1·71	−0·08	−0·41	0·44
30	2·11	−0·22	−0·76	0·20
31	2·07	−0·53	0·55	−0·21
33	1·46	0·00	1·37	0·30
35	2·43	−0·85	−0·76	0·03

Ward and enumeration district number	Component			
	1	2	3	4
St Michael's				
15	1·35	− 0·27	0·55	− 0·30
19	2·23	− 0·31	− 2·06	1·07
20	2·20	0·41	− 1·71	2·18
21	2·36	− 1·14	− 1·66	− 1·03
22	1·72	− 1·18	− 1·21	− 1·81
23	1·39	− 1·09	− 1·06	− 2·17
24	1·85	0·65	− 2·35	0·87
25	3·07	− 1·00	− 0·52	0·36
26	2·68	− 0·77	− 0·06	− 0·75
27	2·42	− 0·92	− 1·06	− 0·68
28	2·00	− 1·07	− 1·56	− 1·60
Southwick				
8	1·15	0·23	0·98	0·26
9	0·20	0·10	1·22	0·18
10	0·00	1·13	0·50	− 0·53
11	− 0·08	1·40	− 0·07	− 1·25
12	0·17	0·92	1·39	− 0·38
13	− 0·88	1·18	0·08	− 2·35
Thornhill				
42	1·22	0·12	0·83	− 0·18
43	1·12	0·09	0·84	− 0·30
44	1·60	0·02	0·29	0·88
45	1·21	0·29	0·20	0·44
46	1·56	− 0·43	1·12	0·48
47	1·35	0·16	0·48	− 0·48
48	1·60	− 0·15	0·65	− 0·16
49	1·46	− 0·20	0·45	− 0·01
50	2·16	− 0·83	− 0·62	− 1·22
51	1·26	− 0·13	1·85	− 0·56
52	2·08	− 0·50	− 0·63	− 0·80
53	1·81	− 0·10	1·14	− 0·11
54	1·93	− 0·73	− 0·06	− 1·07
55	2·29	− 0·89	− 1·42	− 1·01

B] COUNCIL HOUSING AREAS

Ward and enumeration district number	Component			
	1	2	3	4
Bishopwearmouth				
33	1·43	0·92	−0·93	1·25
34	1·33	0·74	0·20	1·37
Bridge				
41	−0·39	0·52	0·10	−1·02
Central				
19	0·27	1·39	−0·93	−0·85
24	−0·41	0·06	0·67	1·04
28	−0·43	0·56	0·65	1·14
30	−0·16	0·23	0·13	−0·92
31	−0·88	−0·43	−0·92	−1·30
Colliery				
54	1·08	−0·73	0·84	0·65
55	0·08	0·38	1·07	0·68
62	−0·49	−0·17	0·43	0·83
63	−0·58	−0·07	0·11	0·75
64	−0·65	−0·73	0·14	0·82
65	0·13	−0·90	0·77	1·01
Hendon				
4	−0·07	0·08	1·39	0·97
5	0·10	−0·66	0·47	1·37
Humbledon				
56	−0·42	−0·71	0·79	2·29
57	−0·21	−0·44	0·72	1·66
58	−0·13	−0·67	0·95	1·97
59	0·17	−0·67	0·74	1·67
60	0·45	−0·60	0·98	1·34
61	0·65	−0·70	1·05	0·10
62	0·22	−0·36	0·77	1·19
63	0·93	−0·99	−0·34	0·22
64	0·09	−0·88	0·57	1·92
65	−0·27	−1·09	0·37	1·36
Hylton Castle				
14	−0·57	−1·15	−0·23	−0·32
15	−0·63	−1·36	−0·09	0·23
16	−0·39	−1·07	0·05	−0·39
17	−0·27	−0·86	−0·19	−1·25
18	−0·33	−0·76	0·17	−1·19
19	−0·20	−0·52	−0·05	−2·07
20	−0·78	−0·93	−0·97	−2·30
21	−0·68	−1·17	−0·41	−1·00
22	−0·38	−1·29	−0·59	−0·98
23	−0·63	−1·08	−0·91	−1·37
24	−0·72	−1·36	−0·81	−1·33
25	−0·23	−0·92	0·42	−1·47
26	−0·33	−1·44	−0·43	−0·01
27	−0·47	−1·60	−0·46	−0·31
28	0·00	−0·13	0·74	−0·06

Ward and enumeration district number	Component			
	1	2	3	4
Monkwearmouth				
1	0·42	0·10	1·55	0·12
9	0·30	0·61	0·58	−0·37
10	−0·23	−0·38	−0·09	−0·39
Pallion				
31	−0·44	−0·14	−0·01	0·18
32	−0·30	−0·70	0·63	0·67
33	0·61	−0·46	2·20	1·41
34	0·50	0·36	1·48	−0·02
36	−0·08	−0·62	1·11	0·49
37	−0·09	−0·78	0·66	1·87
38	0·28	−0·78	1·16	0·94
39	0·59	−0·25	2·16	0·68
Park				
3	−0·73	1·03	0·26	−1·03
5	−0·31	0·70	−0·20	−1·01
Pennywell				
46	−0·98	−0·77	−0·08	1·13
47	−0·25	−0·66	1·09	2·09
48	−0·66	−0·81	0·27	0·77
49	−0·19	−0·89	0·53	0·29
50	−0·07	−1·11	0·48	−0·24
51	−0·54	−1·35	−0·29	1·78
52	−0·44	−0·92	−0·07	1·37
53	−0·50	−1·37	−0·01	1·30
54	−0·45	−1·39	−0·22	1·57
55	−0·44	−1·38	−0·24	0·39
56	−0·07	−0·91	0·13	0·16
57	−0·55	−1·50	−0·32	0·58
58	−0·40	−1·41	−0·83	−0·28
59	−0·44	−1·36	−0·46	1·59
60	−0·59	−1·38	−0·35	−1·81
61	−0·44	−1·40	−0·55	2·00
Roker				
32	1·36	−0·75	0·16	−0·07
34	1·03	−0·26	1·52	0·45
St Michael's				
14	−0·24	−0·11	0·91	1·46
16	0·27	−0·82	0·93	1·56
17	−0·17	−0·96	0·74	1·82
18	1·09	−0·80	0·82	0·81
Southwick				
1	−0·00	−1·28	−0·22	0·71
2	−0·18	−1·32	0·13	0·87
3	0·12	−0·61	0·58	0·77
4	−0·41	−1·26	−0·37	−0·13
5	−0·19	−0·68	0·56	1·05
6	−0·12	−0·75	0·75	0·84
7	0·63	−0·07	1·34	0·33
Thorney Close				
62	−0·16	−1·22	0·20	−0·67

Urban analysis

Ward and enumeration district number	Component			
	1	2	3	4
Thorney Close (*cont.*)				
63	−0·05	−1·11	0·33	0·81
64	−0·27	−0·81	0·46	0·59
65	−0·73	−1·02	−0·76	−1·53
66	−0·46	−0·46	−0·04	−0·70
67	−0·09	−1·08	0·42	−0·02
68	−0·18	−0·59	0·04	0·51
69	−0·60	−1·19	−0·66	1·15
70	−0·33	−1·06	−0·15	0·69
71	−0·36	−0·65	0·52	1·18
72	−0·31	−0·92	0·28	0·52
73	−0·42	−1·07	−0·15	0·62
74	−0·23	−0·88	−0·06	−0·45
75	−0·46	−1·37	−0·21	0·46
76	−0·65	−1·40	−0·57	−0·41
77	−0·54	−1·35	−0·28	0·61
78	−0·28	−1·23	−0·50	−0·61
79	−0·53	−1·23	−0·41	−0·39
80	−0·46	−1·05	−0·50	−1·38
81	−0·91	−1·03	−0·97	−2·24
82	−0·79	−0·85	−0·85	−2·17

NOTES. *i* 'Private housing' enumeration districts include fewer than 25 per cent of council households; 'council housing' enumeration districts include 25 per cent or more council households.

ii The location of wards and enumeration districts is shown in the accompanying map, p. 267.

iii Because of their small population size (see p. 157), a total of nine enumeration districts were excluded from the analysis. Consequently no component scores are recorded for these e.d.s in the above list. The nine excluded e.d.s are: Central Ward, e.d.s 20, 21, 22, 23 and 26 (which comprise the Central Business District); Hendon Ward, e.d. 13 and Monkwearmouth Ward, e.d. 13 (which are areas in which housing had been demolished in 1961); and Hylton Castle Ward, e.d.s 29 and 30 (which, in 1961, were about to be developed as part of a peripheral council estate).

266

MAP OF WARDS AND ENUMERATION DISTRICTS

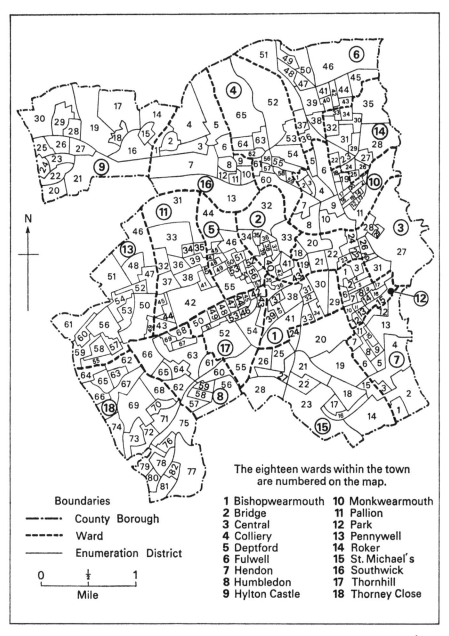

The eighteen wards within the town are numbered on the map.

Boundaries

—·—·— County Borough

------- Ward

——— Enumeration District

0 ½ 1

Mile

1 Bishopwearmouth
2 Bridge
3 Central
4 Colliery
5 Deptford
6 Fulwell
7 Hendon
8 Humbledon
9 Hylton Castle

10 Monkwearmouth
11 Pallion
12 Park
13 Pennywell
14 Roker
15 St. Michael's
16 Southwick
17 Thornhill
18 Thorney Close

AREAL COMBINATIONS OF THE FOUR COMPONENT SCORES

(a) Category as defined by components 1 and 2	(b) Areas	(c) Component 3 score	(d) Component 4 score	(e) Total number of types of combination
1a	15	·	·	
	16	−	·	
	20	·	·	4
	49	·	·	
	60	−	−	
	83	·	−	
1b	17	·	·	
	22	·	+	
	63	−	+	4
	76	·	·	
	61	−	·	
2a	14a	·	−	
	b	·	·	
	21	·	·	
	23	·	·	3
	54	·	·	
	59a	−	−	
	b	·	·	
	75	·	−	
2b	6	·	·	
	13a	·	·	
	b	+	·	
	c	+	·	
	d	+	−	
	19	·	·	
	29	+	·	6
	45	·	·	
	48	+	·	
	58	·	+	
	74a	+	·	
	b	−	·	
	80	·	−	
	84	+	·	
2c	18	−	+	

(a) Category as defined by components 1 and 2	(b) Areas	(c) Component 3 score	(d) Component 4 score	(e) Total number of types of combination
	46	−	.	
	62	−	+	3
	64	−	+	
	72	+	.	
2*d*	65	−	+	1
3*a*	3	.	.	
	12	.	+	
	50	.	+	
	56	.	+	4
	77	+	+	
	81	.	.	
	85	+	.	
3*b*	1	.	.	
	4	+	.	
	10	+	.	
	24	+	.	
	25	+	.	4
	47	+	.	
	52	+	.	
	71	+	.	
	79	.	+	
	86	+	+	
3*c*	5	+	.	
	8	+	.	
	26*a*	+	−	
	b	+	.	
	c	+	.	
	d	.	.	4
	e	+	.	
	32*a*	+	−	
	b	.	.	
	42	.	.	
	67	.	.	
	69	+	−	
3*d*	39	.	+	
	66	−	+	3
	73	.	.	
4*a*	2*a*	.	−	
	b	.	.	
	c	.	+	
	57	.	+	
	88*a*	+	+	4
	b	.	+	
	c	.	−	
	d	.	+	
	e	.	−	
	f	.	+	
4*b*	11	.	.	
	27	.	.	
	32	.	+	

(a) Category as defined by components 1 and 2	(b) Areas	(c) Component 3 score	(d) Component 4 score	(e) Total number of types of combination
	34	.	+	
	36	.	−	5
	51	+	.	
	55	.	+	
	78	.	+	
	82	+	+	
	87	.	.	
	89	.	.	
4c	7a	.	−	
	b	+	.	
	c	.	.	
	d	.	.	
	28	.	.	
	33	.	+	4
	35a	.	+	
	b	.	−	
	38	.	−	
	43	.	.	
	53	.	.	
	68	+	−	
	70	+	.	
4d	9	.	.	
	30	−	−	
	37	−	.	3
	40	−	.	
	42	−	.	
	44	−	.	

NOTES

i   Column (*a*): Categories derived from cross-classification of components 1 and 2. Category 1*a* designates high social class and good housing; 4*d* designates low class and poor housing.

ii   Column (*b*): Area numbers are as shown in Figure 4.11.

iii   Columns (*c*) and (*d*): Scores on components 3 and 4 as follows:

     − designates score below − 1·00

     + designates score above + 1·00

     . designates scores between − 1·00 and + 1·00

iv   Column (*e*): Total number of combinations taking all four components into account.

DISCREPANCIES WITHIN THE MAJOR CATEGORIES

v   in category 2*c*. area 72 has high scores for component 3 whereas the remaining areas have low scores. This is a reflection of the fact that this one area does not display rooming house characteristics even though the selected cutting points for components 1 and 2 include it, marginally, in the rooming-house category. Its scores for components 3 and 4 clearly place the area in a group with the adjacent areas 69 and 71.

vi   in category 2*b*, areas 58 and 74*a* do not conform to the general pattern as regards its scores on components 3 and 4. Area 58 is a mixed area of private and council housing, due to poor delineation of the e.d. boundaries. Area 74*a* has scores on components 3 and 4 which are closer to category 2*a*.

vii in category 4*c*, areas 33 and 35*a* have high scores on component 4 while the general pattern is one of low scores. This appears to be a valid distinction, since both areas include local authority high-rise flats built in this inner East End area during the early 1930s.

viii in category 4*b*, area 36 has low scores on component 4, whereas the remaining areas have high scores. Again, this is an area of central redevelopment, built in more recent years, and the discrepancy reflects the high fertility and young age structure of the area.

ix in category 4*a*, there is a major distinction between the component 4 scores of two types of area: some areas have high scores while others have low scores. This reflects the differing age structure of the various estates within this category which, as is stressed in the text, has important repercussions and is the basis of the major differentials drawn between different council estate areas.

THE QUESTIONNAIRE

---

1   Name.
2   Address.
3   How long have you lived here?
4   Where did you live before that?
5   How many people are there in your family? (Age and education of each, including the parents.)
6   Are there any other people living here with you?
7   Type of house: detached/semi-detached/flat/rooms/other.
8   How many rooms have you in the house?
9   Do you own or rent the house?
10 Age.
11 How old were you when you left school?
12 What sort of school was that?
13 Did you have any full-time or part-time education after you left that school?

} Information for both father and mother

14 Father's occupation.
15 Mother's occupation before marriage.
16 Mother's work now, if any.
17 When exactly is *x* (boy in question) taking the 11-plus examination?
18 Have you given very much thought to this examination?
19 Do you think it is important that he should pass the examination?
20 What sort of school would you like *x* to go to?
21 Is *x* going to have any coaching for the examination apart from the teaching he gets at school?
22 Have you been to the school in the last year to talk to the teachers about *x*?
23 How often have you been and what was it about?
24 What do you think of the school and the teachers there?
25 Do you want *x* to take the G.C.E. examination?
26 What about further education—that is, university or technical college—do you want *x* to have any further education?
27 What exactly were you thinking of?
28 Have you any idea of the sort of job that you would like *x* to get?
29 Do you think that parents should make sacrifices to have their children educated?

30 Do you agree or disagree with this statement: 'The sort of things that are taught in schools are too vague. It would be better for a boy to take a practical course like an apprenticeship than to stay on at school'?

31 Does *x* get any homework from school?

32 Do you help him with his homework, or set him any work yourself?

33 What do you think is the best age for a boy to leave school?

34 How does *x* spend his free time? Is he in any clubs or societies?

35 Has he many friends living close?

36 I should like to know whereabouts your relations live and how often you see them.

Father's parents	Mother's parents
brothers	brothers
sisters	sisters
others close by	others close by

37 Where do your three best friends live?

38 How well would you say that people in this area know each other?

39 How many people in this area do you know by name? Most, some or very few?

40 Would you still live in this area if you had your choice?

41 (If 'no'.) Where would you like to live?

42 Do you think this area is getting better, getting worse or staying the same?

43 What daily newspapers and Sunday papers do you get?

44 Are you a member of any library?

45 Is *x* a member of a library?

46 Which political party did you vote for at the last General Election?

NOTES

i Questionnaire asked of all parents with a son due to sit the local authority 11-plus examination in the following spring.

ii As can be seen, much of the questionnaire was of an informal nature and was supplemented by informal discussion of education and of the local area.

THE SURVEY RESPONDENTS: SOCIAL CLASS AND ATTITUDES TO EDUCATION

THE ALLOCATION OF OCCUPATIONS TO SOCIAL CLASSES

The social class of each family within the survey population was determined on the bases of the father's occupation, the mother's occupation before marriage and the combination of these two. In classifying these occupations, use was made of the General Register Office *Classification of Occupations*[1] in order to provide a standard (although by no means uncriticized) means of allotting occupations to social class categories. Since many of the shipbuilding and heavy industry occupations were of an extremely specialized nature, the official classification proved very useful in distinguishing between skilled and semi-skilled work.

Each occupation was coded and allotted to one of the sixteen socio-economic groups of the classification. These groups are as follows:

1 Employers and managers in central and local government, industry, commerce, etc., in large establishments employing more than twenty-five people.
2 Employers and managers in industry, commerce, etc., in small establishments.
3 Professional workers—self-employed.
4 Professional employees.
5 Intermediate non-manual workers.
6 Junior non-manual workers.
7 Personal service workers.
8 Foremen and supervisors.
9 Skilled manual workers.
10 Semi-skilled manual workers.
11 Unskilled manual workers.
12 Workers on own account (other than professional).
13 Farmers—employers and managers.
14 Farmers—own account.
15 Agricultural workers.
16 Members of the armed services.

[1] *Classification of Occupations, 1960* (H.M.S.O., London, 1960).

For Sunderland respondents, these socio-economic groupings were aggregated into four social classes as follows:

1 upper middle class—socio-economic groups 1, 2, 3, 4.
2 lower middle class—socio-economic groups 5, 6, 7.
3 skilled manual—socio-economic groups 8, 9.
4 semi-skilled and unskilled manual—socio-economic groups 10, 11.

These groups are referred to as 'Social Classes 1, 2, 3 and 4'.

Amongst the survey respondents, there were no occupations falling in socio-economic groups 1, 3, 13, 14, 15 and 16. Only four respondents' occupations fell within group 12 (workers on own account). These were allocated to whichever of the four social classes seemed the most appropriate to the type of work which was done in each case.

Individual difficulties arose in making allocations and two might be specified. First, difficulty arose concerning respondents who were recorded simply as 'miners'. The Registrar General's classification makes a distinction between face workers on one hand and underground and surface workers with less skilled jobs on the other. The respective codings are 010 for face workers (which allots them to socio-economic group 9 and Social Class 3), and 011 for non-face workers (socio-economic group 10, Social Class 4). In cases where full information could be obtained, it was possible to follow this distinction but, where a respondent had been returned as 'miner' (usually by his wife), this was more difficult. It was decided to allot such cases to the non-face-worker category and so place them in the semi-skilled and unskilled social class. In total, this applied to three respondents for whom information was not sufficiently precise. The decision in fact is in accordance with an injunction in the *Classification of Occupations* which suggests that those miners about whom there is insufficient information should be coded as 013 (socio-economic group 10, Social Class 4). This is suggested on a probability basis since the non-face worker category is numerically the larger occupation.

A second difficulty arose over the classification of some of the female occupations. The category 'domestic worker' is classified into socio-economic group 7 (the non-manual service category). On the basis of the aggregation of groups, this would have placed such women into Social Class 2. It was decided that this was so much at variance with the status and type of work involved, that it was best to classify domestic workers together with 'charwomen' who fall within socio-economic group 11 and therefore into the unskilled Social Class 4.

Each father and mother of the respondent families was allotted to one of the four social classes as described above. Tables in the text are presented in terms of three types of classifications of the families: with families grouped in terms of the father's occupation; of the mother's work before marriage; and of the combination of these two—what is referred to as the 'combined class score'. The combined class score is simply the addition of the class scores of both father and mother. The range of class scores is therefore from 1 to 4 in the case of the first two groupings of families and from 2 to 8 in the case of the combined class score. In all cases, lower numbers indicate higher social class.

QUANTIFICATION OF ATTITUDES TOWARDS EDUCATION

Two types of scoring techniques were combined to quantify the attitudes towards education of the parent respondents. The first was a simple weighting of responses; the second was based on the more sophisticated scalogram technique.

Method 1

The simple weighting was based on questions which were shown to be consistently related to the attitude dimension by the scalogram, and the weights were as follows:

	Question	*Response*	*Allotted Score*
20	Kind of school	Grammar school	2
		Technical school	1
		Secondary modern/Don't know	0
22	Visits to school	Yes	1
		No	0
25	G.C.E.	Strongly in favour	2
		Acquiesce	1
		Against/Don't know	0
26	Further education	Strongly in favour	2
		Acquiesce	1
		Against/Don't know	0
32	Help with homework	Yes	1
		No	0
33	Desired leaving age	18-plus	4
		17	3
		16	2
		15	0
45	Library member	Yes	1
		No	0

The possible maximum score was therefore 13, the minimum was 0. The weighting of the responses was determined by a trial-and-error process after experimenting with a number of combinations. The final weights were selected in the light of the accuracy with which they reflected the enthusiasm and interest of the individual parents.

Method 2

The second technique isolated the above seven questions from the fuller range of questions included in the questionnaire and ranked respondents in terms of their consistent pattern of responses. The scalogram technique, introduced by Guttman,[1] represents a considerable refinement of the quantification of attitudes. Most

---

[1] S. A. Stouffer *et al.*, *Studies in social psychology in World War II* : IV, *Measurement and prediction* (Princeton, 1950), chs. 1–9.

methods of quantification are based on a crucial assumption that the various attitude statements which they combine all belong to the same dimension, in other words that they are all measures of the same general field of interest. The scalogram method, however, begins by objectively defining this dimension rather than relying on the assumption of consistency. 'The scalogram hypothesis is that the items have an order such that, ideally, persons who answer a given question favourably all have higher ranks on the score than persons who answer the same score unfavourably.'[1] The items used in scalogram analysis must therefore have a special cumulative property. This 'universe of content' is defined and selected by the 'scaleability' of the responses to the questions, rather than the judgement of the researcher or a panel of experts. Whether or not a particular question belongs to the attitude dimension being measured, is determined by the scalogram technique in terms of the pattern in which respondents' answers arrange themselves. With a perfect scale, one can tell from an individual's total score which items he agreed with and those with which he disagreed. Those questions which do not form a consistent pattern with the majority of responses are excluded from the analysis as being part of a different attitude dimension.

The degree to which responses form a 'scaleable' universe is measured by the 'coefficient of reproducibility', which is 'the empirical frequency with which the values of the attribute do correspond with the proper intervals of the quantitative variable'.[2] This is calculated as follows:

$$1 \cdot 0 - \frac{\text{number of errors}}{\text{number of questions} \times \text{number of respondents}}$$

The coefficient therefore varies between 0·0 and 1·0, with higher values showing greater degrees of reproducibility. Guttman has suggested that coefficients of 0·90 and over may be taken as indicating uni-dimensionality.

The responses to the Sunderland questionnaire were tested by the scalogram method and, in the light of the patterns which emerged, certain of the questions were excluded. Question 30, for example, which tested parents' attitudes towards the content of school courses, showed a pattern which was inconsistent with the majority of responses and was therefore excluded from the analysis. Seven questions, however, did form a scaleable universe and their coefficient of reproducibility was 0·937, which is well within the limit suggested by Guttman.

The scalogram (see pp. 278–82) which is reproduced here shows the manner in which questions and respondents are ranked. Questions are ranked from the most discriminating (i.e. those with fewest favourable responses) to those which are least discriminating (i.e. with most favourable responses). In this case, the most discriminating question was that relating to further education while the least discriminating was that relating to the type of school which parents wished their sons to attend. The ranking of the respondents is made with the objective of achieving the smallest possible number of anomalous responses; those, in other words, which fall outside the cutting points of the scalogram. The Sunderland scalogram does show

---

[1] *Ibid.* p. 9.   [2] *Ibid.* p. 64.

a number of such anomalies in which inconsistent responses exist. Parents thus occasionally can be seen to give favourable responses to questions which are more discriminating than some questions to which they give unfavourable responses. The scores allotted to respondent families, however, are based on consistent patterns of responses and are thus determined by whichever set of cutting points the family lies within. For example, family a7 (from the Deptford survey area) is allotted a score of five even though favourable responses were given to six of the questions, since the favourable reply to the question on further education is inconsistent with the unfavourable reply given to the question of whether the child should have formal education to the age of eighteen or beyond.

In cases such as this, where the pattern of replies was not wholly consistent, families could occasionally have been placed within one of a number of alternative cutting points. In each case, however, the objective of reducing the number of 'errors' was followed. The methods and the definition of the objectives which were adhered to are fully outlined in *Measurement and Prediction*. Suffice it to say that the pattern of scalogram scores was generally consistent and the ranking of questions logical. The relative ranks of the responses concerning parental visits to school, help with homework and the child's membership of libraries, and the fact that these items formed part of a scaleable universe are interesting incidental findings.

The final scores which were used in the analysis were a compound of the two methods outlined, with methods 1 and 2 being combined in a ratio of one to two; the scalogram scores, in other words, being given double weighting. The range of possible scores was therefore from 0 to 29.

Scalogram—attitudes to education

	Responses to questions															
Respondents	1 +	2 +	3 +	4 +	5 +	6 +	7 +	8 +	1 −	2 −	3 −	4 −	5 −	6 −	7 −	8 −
b 22	×	×	×	×	×	×	×	×	.	.	.	.	.	.	.	.
c 3	×	×	×	×	×	×	×	×	.	.	.	.	.	.	.	.
c 4	×	×	×	×	×	×	×	×	.	.	.	.	.	.	.	.
c 9	×	×	×	×	×	×	×	×	.	.	.	.	.	.	.	.
c 14	×	×	×	×	×	×	×	×	.	.	.	.	.	.	.	.
c 15	×	×	×	×	×	×	×	×	.	.	.	.	.	.	.	.
c 17	×	×	×	×	×	×	×	×	.	.	.	.	.	.	.	.
c 19	×	×	×	×	×	×	×	×	.	.	.	.	.	.	.	.
e 2	×	×	×	×	×	×	×	×	.	.	.	.	.	.	.	.
e 5	×	×	×	×	×	×	×	×	.	.	.	.	.	.	.	.
e 7	×	×	×	×	×	×	×	×	.	.	.	.	.	.	.	.
e 10	×	×	×	×	×	×	×	×	.	.	.	.	.	.	.	.
e 14	×	×	×	×	×	×	×	×	.	.	.	.	.	.	.	.
f 2	×	×	×	×	×	×	×	×	.	.	.	.	.	.	.	.
f 16	×	×	×	×	×	×	×	×	.	.	.	.	.	.	.	.
f 17	×	×	×	×	×	×	×	×	.	.	.	.	.	.	.	.
f 24	×	×	×	×	×	×	×	×	.	.	.	.	.	.	.	.
f 26	×	×	×	×	×	×	×	×	.	.	.	.	.	.	.	.

Responses to questions

Respondents	1 +	2 +	3 +	4 +	5 +	6 +	7 +	8 +	1 −	2 −	3 −	4 −	5 −	6 −	7 −	8 −
f 27	×	×	×	×	×	×	×	×	.	.	.	.	.	.	.	.
f 28	×	×	×	×	×	×	×	×	.	.	.	.	.	.	.	.
f 29	×	×	×	×	×	×	×	×	.	.	.	.	.	.	.	.
f 36	×	×	×	×	×	×	×	×	.	.	.	.	.	.	.	.
f 37	×	×	×	×	×	×	×	×	.	.	.	.	.	.	.	.
g 2	×	×	×	×	×	×	×	×	.	.	.	.	.	.	.	.
g 3	×	×	×	×	×	×	×	×	.	.	.	.	.	.	.	.
g 5	×	×	×	×	×	×	×	×	.	.	.	.	.	.	.	.
g 7	×	×	×	×	×	×	×	×	.	.	.	.	.	.	.	.
g 10	×	×	×	×	×	×	×	×	.	.	.	.	.	.	.	.
g 11	×	×	×	×	×	×	×	×	.	.	.	.	.	.	.	.
g 15	×	×	×	×	×	×	×	×	.	.	.	.	.	.	.	.
g 16	×	×	×	×	×	×	×	×	.	.	.	.	.	.	.	.
g 20	×	×	×	×	×	×	×	×	.	.	.	.	.	.	.	.
g 21	×	×	×	×	×	×	×	×	.	.	.	.	.	.	.	.
g 22	×	×	×	×	×	×	×	×	.	.	.	.	.	.	.	.
g 23	×	×	×	×	×	×	×	×	.	.	.	.	.	.	.	.
g 26	×	×	×	×	×	×	×	×	.	.	.	.	.	.	.	.
g 28	×	×	×	×	×	×	×	×	.	.	.	.	.	.	.	.
g 29	×	×	×	×	×	×	×	×	.	.	.	.	.	.	.	.
g 32	×	×	×	×	×	×	×	×	.	.	.	.	.	.	.	.
f 12	×	×	×	×	×	.	×	×	.	.	.	.	.	.	×	.
f 28	×	×	×	×	×	.	×	×	.	.	.	.	.	.	×	.
f 32	×	×	×	×	×	.	×	×	.	.	.	.	.	.	×	.
g 31	×	×	×	×	×	.	×	×	•	.	.	.	.	.	×	.
b 14	×	×	×	×	.	.	×	×	.	.	.	.	×	×	.	.
f 7	×	×	×	×	.	×	×	×	.	.	.	.	×	.	.	.
c 7	×	×	.	×	×	×	×	×	.	.	×	.	.	.	.	.
c 8	×	×	.	×	×	×	×	×	.	.	×	.	.	.	.	.
c 13	×	×	.	×	×	×	×	×	.	.	×	.	.	.	.	.
d 22	×	×	.	×	×	×	×	×	.	.	×	.	.	.	.	.
g 24	×	×	.	×	×	×	×	×	.	.	×	.	.	.	.	.
g 27	×	×	.	×	×	×	×	×	.	.	×	.	.	.	.	.
g 30	×	×	.	×	×	×	×	×	.	.	×	.	.	.	.	.
g 9	×	.	×	×	×	×	×	×	.	×	.	.	.	.	.	.

Allotted score for respondents b 22 to g 9: **8**

Respondents	1 +	2 +	3 +	4 +	5 +	6 +	7 +	8 +	1 −	2 −	3 −	4 −	5 −	6 −	7 −	8 −
a 13	.	×	×	×	×	×	×	×	×	.	.	.	.	.	.	.
b 9	.	×	×	×	×	×	×	×	×	.	.	.	.	.	.	.
b 15	.	×	×	×	×	×	×	×	×	.	.	.	.	.	.	.
c 18	.	×	×	×	×	×	×	×	×	.	.	.	.	.	.	.
d 15	.	×	×	×	×	×	×	×	×	.	.	.	.	.	.	.
e 4	.	×	×	×	×	×	×	×	×	.	.	.	.	.	.	.
f 15	.	×	×	×	×	×	×	×	×	.	.	.	.	.	.	.
f 34	.	×	×	×	×	×	×	×	×	.	.	.	.	.	.	.
f 35	.	×	×	×	×	×	×	×	×	.	.	.	.	.	.	.
g 1	.	×	×	×	×	×	×	×	×	.	.	.	.	.	.	.
c 10	.	×	.	×	×	×	×	×	×	.	×	.	.	.	.	.
f 30	.	×	.	×	×	×	×	×	×	.	×	.	.	.	.	.
g 19	.	×	.	×	×	×	×	×	×	.	×	.	.	.	.	.

Allotted score for respondents a 13 to g 19: **7**

Urban analysis

<table>
<tr><td rowspan="3">Respondents</td><td colspan="16">Responses to questions</td></tr>
<tr><td>1</td><td>2</td><td>3</td><td>4</td><td>5</td><td>6</td><td>7</td><td>8</td><td>1</td><td>2</td><td>3</td><td>4</td><td>5</td><td>6</td><td>7</td><td>8</td></tr>
<tr><td>+</td><td>+</td><td>+</td><td>+</td><td>+</td><td>+</td><td>+</td><td>+</td><td>−</td><td>−</td><td>−</td><td>−</td><td>−</td><td>−</td><td>−</td><td>−</td></tr>
<tr><td>b 19</td><td>.</td><td>.</td><td>×</td><td>×</td><td>×</td><td>×</td><td>×</td><td>×</td><td>×</td><td>×</td><td>.</td><td>.</td><td>.</td><td>.</td><td>.</td><td>.</td></tr>
<tr><td>b 24</td><td>.</td><td>.</td><td>×</td><td>×</td><td>×</td><td>×</td><td>×</td><td>×</td><td>×</td><td>×</td><td>.</td><td>.</td><td>.</td><td>.</td><td>.</td><td>.</td></tr>
<tr><td>b 32</td><td>.</td><td>.</td><td>×</td><td>×</td><td>×</td><td>×</td><td>×</td><td>×</td><td>×</td><td>×</td><td>.</td><td>.</td><td>.</td><td>.</td><td>.</td><td>.</td></tr>
<tr><td>c 2</td><td>.</td><td>.</td><td>×</td><td>×</td><td>×</td><td>×</td><td>×</td><td>×</td><td>×</td><td>×</td><td>.</td><td>.</td><td>.</td><td>.</td><td>.</td><td>.</td></tr>
<tr><td>c 6</td><td>.</td><td>.</td><td>×</td><td>×</td><td>×</td><td>×</td><td>×</td><td>×</td><td>×</td><td>×</td><td>.</td><td>.</td><td>.</td><td>.</td><td>.</td><td>.</td></tr>
<tr><td>d 28</td><td>.</td><td>.</td><td>×</td><td>×</td><td>×</td><td>×</td><td>×</td><td>×</td><td>×</td><td>×</td><td>.</td><td>.</td><td>.</td><td>.</td><td>.</td><td>.</td></tr>
<tr><td>e 11</td><td>.</td><td>.</td><td>×</td><td>×</td><td>×</td><td>×</td><td>×</td><td>×</td><td>×</td><td>×</td><td>.</td><td>.</td><td>.</td><td>.</td><td>.</td><td>.</td></tr>
<tr><td>e 13</td><td>.</td><td>.</td><td>×</td><td>×</td><td>×</td><td>×</td><td>×</td><td>×</td><td>×</td><td>×</td><td>.</td><td>.</td><td>.</td><td>.</td><td>.</td><td>.</td></tr>
<tr><td>f 4</td><td>.</td><td>.</td><td>×</td><td>×</td><td>×</td><td>×</td><td>×</td><td>×</td><td>×</td><td>×</td><td>.</td><td>.</td><td>.</td><td>.</td><td>.</td><td>.</td></tr>
<tr><td>f 13</td><td>.</td><td>.</td><td>×</td><td>×</td><td>×</td><td>×</td><td>×</td><td>×</td><td>×</td><td>×</td><td>.</td><td>.</td><td>.</td><td>.</td><td>.</td><td>.</td></tr>
<tr><td>f 19</td><td>.</td><td>.</td><td>×</td><td>×</td><td>×</td><td>×</td><td>×</td><td>×</td><td>×</td><td>×</td><td>.</td><td>.</td><td>.</td><td>.</td><td>.</td><td>.</td></tr>
<tr><td>f 20</td><td>.</td><td>.</td><td>×</td><td>×</td><td>×</td><td>×</td><td>×</td><td>×</td><td>×</td><td>×</td><td>.</td><td>.</td><td>.</td><td>.</td><td>.</td><td>.</td></tr>
<tr><td>g 4</td><td>.</td><td>.</td><td>×</td><td>×</td><td>×</td><td>×</td><td>×</td><td>×</td><td>×</td><td>×</td><td>.</td><td>.</td><td>.</td><td>.</td><td>.</td><td>.</td></tr>
<tr><td>g 25</td><td>.</td><td>.</td><td>×</td><td>×</td><td>×</td><td>×</td><td>×</td><td>×</td><td>×</td><td>×</td><td>.</td><td>.</td><td>.</td><td>.</td><td>.</td><td>.</td></tr>
<tr><td>e 1</td><td>×</td><td>.</td><td>×</td><td>×</td><td>.</td><td>×</td><td>×</td><td>×</td><td>.</td><td>×</td><td>.</td><td>.</td><td>×</td><td>.</td><td>.</td><td>.</td></tr>
<tr><td>d 23</td><td>.</td><td>.</td><td>×</td><td>×</td><td>×</td><td>×</td><td>×</td><td>.</td><td>×</td><td>×</td><td>.</td><td>.</td><td>.</td><td>.</td><td>.</td><td>×</td></tr>
<tr><td>f 21</td><td>.</td><td>.</td><td>×</td><td>×</td><td>×</td><td>×</td><td>.</td><td>×</td><td>×</td><td>×</td><td>.</td><td>.</td><td>.</td><td>.</td><td>×</td><td>.</td></tr>
<tr><td>d 26</td><td>.</td><td>.</td><td>×</td><td>×</td><td>×</td><td>.</td><td>×</td><td>×</td><td>×</td><td>×</td><td>.</td><td>.</td><td>.</td><td>×</td><td>.</td><td>.</td></tr>
<tr><td>f 9</td><td>.</td><td>.</td><td>×</td><td>×</td><td>×</td><td>.</td><td>×</td><td>×</td><td>×</td><td>×</td><td>.</td><td>.</td><td>.</td><td>×</td><td>.</td><td>.</td></tr>
<tr><td>d 20</td><td>.</td><td>.</td><td>×</td><td>×</td><td>.</td><td>.</td><td>×</td><td>×</td><td>×</td><td>×</td><td>.</td><td>.</td><td>×</td><td>×</td><td>.</td><td>.</td></tr>
<tr><td>f 22</td><td>.</td><td>.</td><td>×</td><td>×</td><td>.</td><td>.</td><td>×</td><td>×</td><td>×</td><td>×</td><td>.</td><td>.</td><td>×</td><td>×</td><td>.</td><td>.</td></tr>
<tr><td>g 17</td><td>.</td><td>.</td><td>×</td><td>.</td><td>×</td><td>×</td><td>×</td><td>×</td><td>×</td><td>×</td><td>.</td><td>×</td><td>.</td><td>.</td><td>.</td><td>.</td></tr>
</table>

Allotted score for respondents b 19 to g 17: **6**

<table>
<tr><td>a 3</td><td>.</td><td>.</td><td>.</td><td>×</td><td>×</td><td>×</td><td>×</td><td>×</td><td>×</td><td>×</td><td>×</td><td>.</td><td>.</td><td>.</td><td>.</td><td>.</td></tr>
<tr><td>b 30</td><td>.</td><td>.</td><td>.</td><td>×</td><td>×</td><td>×</td><td>×</td><td>×</td><td>×</td><td>×</td><td>×</td><td>.</td><td>.</td><td>.</td><td>.</td><td>.</td></tr>
<tr><td>c 12</td><td>.</td><td>.</td><td>.</td><td>×</td><td>×</td><td>×</td><td>×</td><td>×</td><td>×</td><td>×</td><td>×</td><td>.</td><td>.</td><td>.</td><td>.</td><td>.</td></tr>
<tr><td>d 14</td><td>.</td><td>.</td><td>.</td><td>×</td><td>×</td><td>×</td><td>×</td><td>×</td><td>×</td><td>×</td><td>×</td><td>.</td><td>.</td><td>.</td><td>.</td><td>.</td></tr>
<tr><td>d 16</td><td>.</td><td>.</td><td>.</td><td>×</td><td>×</td><td>×</td><td>×</td><td>×</td><td>×</td><td>×</td><td>×</td><td>.</td><td>.</td><td>.</td><td>.</td><td>.</td></tr>
<tr><td>e 8</td><td>.</td><td>.</td><td>.</td><td>×</td><td>×</td><td>×</td><td>×</td><td>×</td><td>×</td><td>×</td><td>×</td><td>.</td><td>.</td><td>.</td><td>.</td><td>.</td></tr>
<tr><td>e 9</td><td>.</td><td>.</td><td>.</td><td>×</td><td>×</td><td>×</td><td>×</td><td>×</td><td>×</td><td>×</td><td>×</td><td>.</td><td>.</td><td>.</td><td>.</td><td>.</td></tr>
<tr><td>e 12</td><td>.</td><td>.</td><td>.</td><td>×</td><td>×</td><td>×</td><td>×</td><td>×</td><td>×</td><td>×</td><td>×</td><td>.</td><td>.</td><td>.</td><td>.</td><td>.</td></tr>
<tr><td>f 8</td><td>.</td><td>.</td><td>.</td><td>×</td><td>×</td><td>×</td><td>×</td><td>×</td><td>×</td><td>×</td><td>×</td><td>.</td><td>.</td><td>.</td><td>.</td><td>.</td></tr>
<tr><td>f 11</td><td>.</td><td>.</td><td>.</td><td>×</td><td>×</td><td>×</td><td>×</td><td>×</td><td>×</td><td>×</td><td>×</td><td>.</td><td>.</td><td>.</td><td>.</td><td>.</td></tr>
<tr><td>f 33</td><td>.</td><td>.</td><td>.</td><td>×</td><td>×</td><td>×</td><td>×</td><td>×</td><td>×</td><td>×</td><td>×</td><td>.</td><td>.</td><td>.</td><td>.</td><td>.</td></tr>
<tr><td>g 12</td><td>.</td><td>.</td><td>.</td><td>×</td><td>×</td><td>×</td><td>×</td><td>×</td><td>×</td><td>×</td><td>×</td><td>.</td><td>.</td><td>.</td><td>.</td><td>.</td></tr>
<tr><td>g 18</td><td>.</td><td>.</td><td>.</td><td>×</td><td>×</td><td>×</td><td>×</td><td>×</td><td>×</td><td>×</td><td>×</td><td>.</td><td>.</td><td>.</td><td>.</td><td>.</td></tr>
<tr><td>a 7</td><td>×</td><td>.</td><td>.</td><td>×</td><td>×</td><td>×</td><td>×</td><td>×</td><td>.</td><td>×</td><td>×</td><td>.</td><td>.</td><td>.</td><td>.</td><td>.</td></tr>
<tr><td>d 12</td><td>×</td><td>.</td><td>.</td><td>×</td><td>×</td><td>×</td><td>×</td><td>×</td><td>.</td><td>×</td><td>×</td><td>.</td><td>.</td><td>.</td><td>.</td><td>.</td></tr>
<tr><td>b 6</td><td>×</td><td>.</td><td>.</td><td>×</td><td>×</td><td>.</td><td>×</td><td>×</td><td>.</td><td>×</td><td>×</td><td>.</td><td>.</td><td>×</td><td>.</td><td>.</td></tr>
<tr><td>c 11</td><td>.</td><td>×</td><td>.</td><td>×</td><td>×</td><td>×</td><td>×</td><td>×</td><td>×</td><td>.</td><td>×</td><td>.</td><td>.</td><td>.</td><td>.</td><td>.</td></tr>
<tr><td>d 32</td><td>.</td><td>×</td><td>.</td><td>×</td><td>×</td><td>×</td><td>×</td><td>×</td><td>×</td><td>.</td><td>×</td><td>.</td><td>.</td><td>.</td><td>.</td><td>.</td></tr>
<tr><td>d 6</td><td>.</td><td>.</td><td>.</td><td>×</td><td>×</td><td>×</td><td>×</td><td>.</td><td>×</td><td>×</td><td>×</td><td>.</td><td>.</td><td>.</td><td>.</td><td>×</td></tr>
<tr><td>b 34</td><td>.</td><td>.</td><td>.</td><td>×</td><td>×</td><td>×</td><td>.</td><td>×</td><td>×</td><td>×</td><td>×</td><td>.</td><td>.</td><td>×</td><td>.</td><td>.</td></tr>
<tr><td>f 31</td><td>.</td><td>.</td><td>.</td><td>×</td><td>×</td><td>×</td><td>.</td><td>×</td><td>×</td><td>×</td><td>×</td><td>.</td><td>.</td><td>×</td><td>.</td><td>.</td></tr>
<tr><td>a 10</td><td>.</td><td>.</td><td>.</td><td>×</td><td>×</td><td>.</td><td>×</td><td>×</td><td>×</td><td>×</td><td>×</td><td>.</td><td>.</td><td>×</td><td>.</td><td>.</td></tr>
<tr><td>b 37</td><td>.</td><td>.</td><td>.</td><td>×</td><td>×</td><td>.</td><td>×</td><td>×</td><td>×</td><td>×</td><td>×</td><td>.</td><td>.</td><td>×</td><td>.</td><td>.</td></tr>
<tr><td>f 5</td><td>.</td><td>.</td><td>.</td><td>×</td><td>×</td><td>.</td><td>×</td><td>×</td><td>×</td><td>×</td><td>×</td><td>.</td><td>.</td><td>×</td><td>.</td><td>.</td></tr>
<tr><td>f 6</td><td>.</td><td>.</td><td>.</td><td>×</td><td>×</td><td>.</td><td>×</td><td>×</td><td>×</td><td>×</td><td>×</td><td>.</td><td>.</td><td>×</td><td>.</td><td>.</td></tr>
<tr><td>a 15</td><td>.</td><td>.</td><td>.</td><td>×</td><td>.</td><td>×</td><td>×</td><td>×</td><td>×</td><td>×</td><td>×</td><td>.</td><td>×</td><td>.</td><td>.</td><td>.</td></tr>
<tr><td>d 25</td><td>.</td><td>.</td><td>.</td><td>×</td><td>.</td><td>×</td><td>×</td><td>×</td><td>×</td><td>×</td><td>×</td><td>.</td><td>×</td><td>.</td><td>.</td><td>.</td></tr>
</table>

Allotted score for respondents a 3 to d 25: **5**

Responses to questions

Respondents	1 +	2 +	3 +	4 +	5 +	6 +	7 +	8 +	1 −	2 −	3 −	4 −	5 −	6 −	7 −	8 −
b 3	.	.	.	.	×	×	×	×	×	×	×	×	.	.	.	.
b 28	.	.	.	.	×	×	×	×	×	×	×	×	.	.	.	.
b 29	.	.	.	.	×	×	×	×	×	×	×	×	.	.	.	.
b 31	.	.	.	.	×	×	×	×	×	×	×	×	.	.	.	.
e 3	.	.	.	.	×	×	×	×	×	×	×	×	.	.	.	.
f 10	.	.	.	.	×	×	.	×	×	×	×	×	.	.	×	.
d 13	.	×	.	.	×	×	×	×	×	.	×	×	.	.	.	.
a 6	.	.	×	.	×	×	×	×	×	×	.	×	.	.	.	.
g 14	.	.	×	.	×	×	×	×	×	×	.	×	.	.	.	.
a 2	.	×	.	.	×	×	×	.	×	.	×	×	.	.	.	×
c 1	.	.	×	.	×	×	.	×	×	×	.	×	.	.	×	.
e 6	.	.	×	.	×	×	.	×	×	×	.	×	.	.	×	.
f 23	.	.	×	.	×	.	×	×	×	×	.	×	.	×	.	.
f 1	.	.	×	.	×	×	.	.	×	×	.	×	.	.	×	×

Allotted score for respondents b 3 to f 1: **4**

Respondents	1 +	2 +	3 +	4 +	5 +	6 +	7 +	8 +	1 −	2 −	3 −	4 −	5 −	6 −	7 −	8 −
b 13	.	.	.	.	.	×	×	×	×	×	×	×	×	.	.	.
b 18	.	.	.	.	.	×	×	×	×	×	×	×	×	.	.	.
b 20	.	.	.	.	.	×	×	×	×	×	×	×	×	.	.	.
d 5	.	.	.	.	.	×	×	×	×	×	×	×	×	.	.	.
b 12	.	.	×	.	.	×	×	×	×	×	.	×	×	.	.	.
d 8	.	.	×	.	.	×	×	×	×	×	.	×	×	.	.	.
d 18	.	.	×	.	.	×	×	×	×	×	.	×	×	.	.	.
f 14	.	.	×	.	.	×	×	×	×	×	.	×	×	.	.	.
b 4	.	.	.	×	.	×	×	.	×	×	×	.	×	.	.	×
b 11	.	.	.	×	.	×	×	.	×	×	×	.	×	.	.	×
g 13	.	.	.	×	.	×	.	×	×	×	×	.	×	.	×	.
d 35	.	.	.	.	.	×	×	.	×	×	×	×	×	.	.	×

Allotted score for respondents b 13 to d 35: **3**

Respondents	1 +	2 +	3 +	4 +	5 +	6 +	7 +	8 +	1 −	2 −	3 −	4 −	5 −	6 −	7 −	8 −
a 4	.	.	.	.	.	.	×	×	×	×	×	×	×	×	.	.
a 17	.	.	.	.	.	.	×	×	×	×	×	×	×	×	.	.
b 10	.	.	.	.	.	.	×	×	×	×	×	×	×	×	.	.
b 25	.	.	.	.	.	.	×	×	×	×	×	×	×	×	.	.
d 1	.	.	.	.	.	.	×	×	×	×	×	×	×	×	.	.
d 4	.	.	.	.	.	.	×	×	×	×	×	×	×	×	.	.
d 19	.	.	.	.	.	.	×	×	×	×	×	×	×	×	.	.
d 29	.	.	.	.	.	.	×	×	×	×	×	×	×	×	.	.
d 11	.	.	×	.	.	.	×	×	×	×	.	×	×	×	.	.
a 1	.	.	.	×	.	.	×	×	×	×	×	.	×	×	.	.
d 2	.	.	.	×	.	.	×	×	×	×	×	.	×	×	.	.
d 17	.	.	.	×	.	.	×	×	×	×	×	.	×	×	.	.
d 10	.	.	.	.	×	.	×	×	×	×	×	×	.	×	.	.

Allotted score for respondents a 4 to d 10: **2**

Respondents	1 +	2 +	3 +	4 +	5 +	6 +	7 +	8 +	1 −	2 −	3 −	4 −	5 −	6 −	7 −	8 −
a 14	.	.	.	.	.	.	.	×	×	×	×	×	×	×	×	.
b 1	.	.	.	.	.	.	.	×	×	×	×	×	×	×	×	.
b 7	.	.	.	.	.	.	.	×	×	×	×	×	×	×	×	.
d 3	.	.	.	.	.	.	.	×	×	×	×	×	×	×	×	.
d 21	.	.	.	.	.	.	.	×	×	×	×	×	×	×	×	.
d 30	.	.	.	.	.	.	.	×	×	×	×	×	×	×	×	.
d 33	.	.	.	.	.	.	.	×	×	×	×	×	×	×	×	.

Responses to questions

Respondents	1 +	2 +	3 +	4 +	5 +	6 +	7 +	8 +	1 −	2 −	3 −	4 −	5 −	6 −	7 −	8 −
d 34	.	.	.	.	.	.	.	×	×	×	×	×	×	×	×	.
a 8	.	.	×	.	.	.	.	×	×	×	.	×	×	×	×	.
b 2	.	.	×	.	.	.	.	×	×	×	.	×	×	×	×	.
f 3	.	.	.	×	.	.	.	×	×	×	×	.	×	×	×	.
a 11	.	.	.	.	×	.	.	×	×	×	×	×	.	×	×	.
b 36	.	.	.	.	×	.	.	×	×	×	×	×	.	×	×	.
d 27	.	.	.	.	×	.	.	×	×	×	×	×	.	×	×	.
b 26	.	.	.	.	.	×	.	×	×	×	×	×	×	.	×	.
b 33	.	.	.	.	.	×	.	×	×	×	×	×	×	.	×	.

Allotted score for respondents a 14 to b 33: 1

Respondents	1 +	2 +	3 +	4 +	5 +	6 +	7 +	8 +	1 −	2 −	3 −	4 −	5 −	6 −	7 −	8 −
b 23	.	.	×	.	.	.	.	.	×	×	.	×	×	×	×	×
a 16	.	.	.	.	×	×	.	.	×	×	×	×	.	.	×	×
b 8	.	.	.	.	×	×	.	.	×	×	×	×	.	.	×	×
b 27	.	.	.	.	×	.	.	.	×	×	×	×	.	×	×	×
a 5	.	.	.	.	.	×	.	.	×	×	×	×	×	.	×	×
a 12	.	.	.	.	.	×	.	.	×	×	×	×	×	.	×	×
d 7	.	.	.	.	.	×	.	.	×	×	×	×	×	.	×	×
d 31	.	.	.	.	.	×	.	.	×	×	×	×	×	.	×	×
c 5	.	.	.	.	.	.	×	.	×	×	×	×	×	×	.	.
a 9	.	.	.	.	.	.	.	.	×	×	×	×	×	×	×	×
b 5	.	.	.	.	.	.	.	.	×	×	×	×	×	×	×	×
b 21	.	.	.	.	.	.	.	.	×	×	×	×	×	×	×	×
b 35	.	.	.	.	.	.	.	.	×	×	×	×	×	×	×	×
c 16	.	.	.	.	.	.	.	.	×	·×	×	×	×	×	×	×
d 6	.	.	.	.	.	.	.	.	×	×	×	×	×	×	×	×
d 24	.	.	.	.	.	.	.	.	×	×	×	×	×	×	×	×
d 36	.	.	.	.	.	.	.	.	×	×	×	×	×	×	×	×

Allotted score for respondents b 23 to d 36: 0

NOTES

i Crosses record the (dichotomized) responses to eight questions. Favourable responses lie to the left, unfavourable to the right of the centre. Respondents are ranked in order of favourable attitudes to education.

ii The questions (relating to sons of parent respondents) are as follows:
1 Desire for further education (+ yes, − no).
2 Desire to stay at school beyond age 18 (+ yes, − no).
3 Member of library (+ yes, − no).
4 Desire to sit G.C.E. (+ yes, − no).
5 Help with homework (+ yes, − no).
6 Visit school (+ yes, − no).
7 Desire to stay at school beyond age 16 (+ yes, − no).
8 Kind of school desired (+ grammar or technical, − secondary modern).

iii Respondents' area of residence is recorded as follows:
a—Deptford; b—Upper Hendon; c—Lower Hendon; d—Thorney Close; e—Bishopwearmouth; f—Thornhill; g—Alexandra Road.

BIBLIOGRAPHY

The principal primary sources of data used in this study are as follows:

(i) General Register Office, *Census of England and Wales* (various volumes, 1801–1961); Census enumeration district material (scales A and D), Sunderland C.B., 1961 (supplied privately).

(ii) *Assessment for the relief of the poor of the townships of Bishopwearmouth and Monkwearmouth and the parish of Sunderland* (3 vols. 1850); *Valuation Assessment, 1892* (10 vols. 1892); *Sunderland valuation list, 1963* (7 vols. 1963).

(iii) *Register of electors, Sunderland County Borough*, 1963.

(iv) *Ward's directory of the County Borough of Sunderland* (London, various dates).

(v) *Results of 11-plus examinations, 1962 and 1963* (Sunderland Education Office).

The following list is of references explicitly consulted in writing the text. For more comprehensive bibliographies and surveys of the literature on urban ecology and urban analysis, the following works are recommended: G. A. Theodorson (ed.), *Studies in human ecology* (Evanston, Illinois, 1961); J. A. Quinn, 'Topical summary of current literature on human ecology', *Am. J. Sociol.* XLVI (1940), 191–226; P. K. Hatt and A. J. Reiss (eds.), *Cities and society : the revised reader in urban sociology* (2nd ed. Glencoe, Illinois, 1957); H. M. Mayer and C. F. Kohn (eds.), *Readings in urban geography* (Chicago, 1959); P. M. Hauser and L. F. Schnore (eds.), *The study of urbanization* (New York, 1965); and R. Glass, 'Urban sociology in Great Britain: a trend report', *Current Sociology*, IV (1955). The literature on education is best covered in J. Floud, 'The sociology of education: a trend report', *Current Sociology*, VII (1958), and in A. H. Halsey, J. Floud and C. A. Anderson (eds.), *Education, economy and society : a reader in the sociology of education* (New York, 1961).

With the revival of interest in the ideas of the early human ecologists, a number of their writings have recently been reprinted in three of the above works. To facilitate easy access to these articles, references to reprints are added as suffixes, where appropriate, as follows: Theodorson, 1961; Hatt and Reiss, 1957; Mayer and Kohn, 1959.

LIST OF REFERENCES

ABBOTT, E. *The tenements of Chicago, 1908–1935* (Chicago, 1936).

ACKERMAN, E. A. *Geography as a fundamental research discipline*. University of Chicago, Department of Geography Research Paper no. 53 (Chicago, 1958).

'Geography and demography', in P. M. Hauser and O. D. Duncan (eds.), *The study of population* (Chicago, 1959), pp. 717–27.

'Where is a research frontier?', *Ann. Ass. Am. Geogr.* LIII (1963), 429–40.

AHMAD, Q. *Indian cities: characteristics and correlates*. University of Chicago, Department of Geography Research Paper no. 102 (Chicago, 1965).

ALIHAN, M. A. *Social ecology: a critical analysis* (New York, 1938, reprinted 1964).

ALONSO, W. 'The historical and the structural theories of urban form', *Land Economics*, XL (1964), 227–31.

ANDERSON, T. R. and EGELAND, J. A. 'Spatial aspects of social area analysis', *Am. Sociol. Rev.* XXVI (1961), 392–8.

AXELROD, M. 'Urban structure and social participation', *Am. Sociol. Rev.* XXI (1956), 13–18.

BARLOW, J. E. and RAMSDALE, G. I. 'Balanced population: an experiment at Silksworth, overspill township for Sunderland', *J. Town Planning Inst.* LII (1966), 265–9.

BARROWS, H. H. 'Geography as human ecology', *Ann. Ass. Am. Geogr.* XIII (1923), 1–14.

BELL, W. 'Economic, family and ethnic status: an empirical test', *Am. Sociol. Rev.* XX (1955), 45–52.

BELL, W. and FORCE, M. T. 'Urban neighbourhood types and participation in formal associations', *Am. Sociol. Rev.* XXI (1956), 25–34.

BERNSTEIN, B. 'Language and social class', *Br. J. Sociol.* XI (1960), 271–6.

BERRY, B. J. L. 'A note concerning methods of classification', *Ann. Ass. Am. Geogr.* XLVIII (1958), 300–3.

'An inductive approach to the regionalization of economic development' in N. Ginsburg (ed.), *Essays on geography and economic development* (Chicago, 1960), pp. 78–107.

'A method for deriving multi-factor uniform regions', *Przegl. Geogr.* XXXIII (1961), 263–82.

'Research frontiers in urban geography', in P. M. Hauser and L. F. Schnore (eds.), *The study of urbanization* (New York, 1965), pp. 403–30.

Essays on commodity flows and the spatial structure of the Indian economy. University of Chicago, Department of Geography Research Paper no. 111 (Chicago, 1966).

284

BESHERS, J. M. *Urban social structure* (New York, 1962).

BEWS, J. W. *Human ecology* (London, 1935).

BOARD OF TRADE. *The North-East: a programme for regional development and growth* (Cmnd. 2206, London, 1963).

BOGUE, D. J. *State economic areas: a description of the procedures used in making a functional grouping of the counties of the United States* (Washington, 1951).

'Population distribution', in P. M. Hauser and O. D. Duncan (eds.), *The study of population* (Chicago, 1959), pp. 383–99.

BOGUE, D. J. and HARRIS, D. L. *Comparative population and urban research via multiple regression and covariance analysis* (Oxford, Ohio, 1954).

BOSSARD, J. H. S. 'Residential propinquity as a factor in marriage selection', *Am. J. Sociol.* XXXVIII (1932), 219–24.

BOTT, E. 'The concept of class as a reference group', *Human Relations*, III (1954), 259–85.

Family and social network: roles, norms and external relations in ordinary urban families (London, 1957).

BRIGGS, A. *Victorian Cities* (London, 1963).

BUNGE, W. 'Gerrymandering, geography and grouping', *Geogr. Rev.* LVI (1966), 256–63.

BURGESS, D. A. 'Some aspects of the geography of the ports of Sunderland, Seaham, and the Hartlepools', unpublished M.A. thesis, University of Durham, 1961.

BURGESS, E. W. 'The growth of the city: an introduction to a research project', *Proc. and Pap. Am. Sociol. Soc.* XVIII (1924), 85–97. (Also in Theodorson, 1961.)

(ed.). *The urban community* (Chicago, 1925).

BURGESS, E. W. and BOGUE, D. J. (eds.). *Contributions to urban sociology* (Chicago, 1964).

BURNETT, J. *History of the town and port of Sunderland and the parishes of Bishop-wearmouth and Monkwearmouth* (Sunderland, 1830).

CAMPBELL, W. J. 'The influence of the home environment on educational progress in secondary schools', *Br. J. Educ. Psychol.* XXII (1952), 89–100.

CAPLOW, T. 'Urban structure in France', *Am. Sociol. Rev.* XVII (1952), 544–9. (Also in Theodorson, 1961.)

CAPLOW, T. and FORMAN, R. 'Neighbourhood interaction in a homogeneous community', *Am. Sociol. Rev.* XV (1950), 357–67.

CAREY, G. W. 'The regional interpretation of Manhattan population and housing patterns through factor analysis', *Geogr. Rev.* LVI (1966), 551–69.

CARTER, H. 'Aberystwyth: the modern development of a medieval castle town in Wales', *Trans. Inst. Br. Geogr.* XXV (1958), 239–53.

CATTELL, R. B. *Factor analysis: an introduction and manual for the psychologist and social scientist* (New York, 1952).

CENTERS, R. *The psychology of social classes* (Princeton, 1949).

CENTRE FOR URBAN STUDIES (University College, London). 'A note on the principal component analysis of 1961 enumeration district data for London Administrative County', unpublished report, no date.

Bibliography

CHAPIN, F. S. 'The psychology of housing', *Social Forces*, XXX (1951), 11–15.

CHILD, C. M. 'Biological foundations of social integration', *Proc. Pap. Am. Sociol. Soc.* XXII (1928), 26–42.

CLEMENTS, F. E. *Plant succession: an analysis of the development of vegetation* (Washington, D.C., 1916).

COLBY, C. C. 'Centrifugal and centripetal forces in urban geography', *An. Ass. Am. Geogr.* XXIII (1933), 1–20. (Also in Mayer and Kohn, 1959.)

COLLISON, P. 'Occupation, education and housing in an English city', *Am. J. Sociol.* LXV (1960), 588–97.

COLLISON, P. and MOGEY, J. M. 'Residence and social class in Oxford', *Am. J. Sociol.* LXIV (1959), 599–605.

COMART RESEARCH LTD. *Survey of incomes and households in the United Kingdom 1964* (London, 1964).

COOLEY, W. W. and LOHNES, P. R. *Multivariate procedures for the behavioural sciences* (New York, 1962).

CULLINGWORTH, J. B. *Housing in transition: a case study in the city of Lancaster, 1958–1962* (London, 1963).

DAVIE, M. R. 'The pattern of urban growth', in G. P. Murdock (ed.), *Studies in the science of society* (New Haven, 1937), pp. 131–61. (Also in Theodorson, 1961.)

DAVIE, M. R. and REEVES, R. J. 'Propinquity of residence before marriage', *Am. J. Sociol.* XLIV (1939), 510–17.

DAVIS, A. 'Motivation of the underprivileged worker', in W. F. Whyte (ed.), *Industry and society* (New York, 1946), pp. 84–106.
Social class influences upon learning (Cambridge, Mass., 1948).

DICKINSON, R. E. *City region and regionalism: a geographical contribution to human ecology* (London, 1947).

DOUGLAS, J. W. B. *The home and the school: a study of ability and attainment in the primary school* (London, 1964).

DUNCAN, B. 'Variables in urban morphology', in E. W. Burgess and D. J. Bogue (eds.), *Contributions to urban sociology* (Chicago, 1964), pp. 17–30.

DUNCAN, O. D. 'Review' (of Shevky and Bell, 1955), *Am. J. Sociol.* LVI (1955), 84–5.
'Human ecology and population studies', in P. M. Hauser and O. D. Duncan (eds.), *The study of population* (Chicago, 1959), pp. 678–716.

DUNCAN, O. D., CUZZORT, R. P. and DUNCAN, B. *Statistical geography: problems in analyzing areal data* (Glencoe, Illinois, 1961).

DUNCAN, O. D. and DAVIS, B. 'An alternative to ecological correlations', *Am. Sociol. Rev.* XVIII (1953), 665–6.

DUNCAN, O. D. and DUNCAN, B. 'Residential distribution and occupational stratification', *Am. J. Sociol.* LX (1955), 493–506.

DUNCAN, O. D. and REISS, A. J. Jr. *Social characteristics of urban and rural communities, 1950* (New York, 1956).

DUNCAN, O. D. and SCHNORE, L. F. 'Cultural, behavioural and ecological perspectives in the study of social behaviour', *Am. J. Sociol.* LXV (1959), 132–46.

EUGENICS REVIEW. 'Fertility differentials in England and Wales: some facts', *Eugenics Review*, LIX (1967), 70–2.

EYRE, S. R. 'Determinism and the ecological approach to geography', *Geography*, XLIX (1964), 369–76.

FESTINGER, L., SCHACHTER, S. and BACK, K. *Social pressures in informal groups* (New York, 1950).

FIREY, W. 'Sentiment and symbolism as ecological variables', *Am. Sociol. Rev.* X (1945), 140–8. (Also in Theodorson, 1961.)
Land use in central Boston (Cambridge, Mass., 1947).

FLEMING, C. M. *The social psychology of education* (London, 1944).

FLOUD, J. 'Education and social class in the welfare state', in A. V. Judges (ed.), *Looking forward in education* (London, 1955), pp. 38–59.

FLOUD, J., HALSEY, A. H. and MARTIN, F. M. *Social class and educational opportunity* (London, 1956).

FOLEY, D. L. 'Census tracts and urban research', *J. Am. statist. Ass.* XLVIII (1953), 733–42.

FORDYCE, W. *The history and antiquities of the County Palatine of Durham* (Newcastle, 1857).

FORM, W. H. 'Stratification in low and middle income housing areas', *J. Social Issues*, VII (1951), 109–31.

FORM, W. H., SMITH, J. *et al.* 'The comparability of alternative approaches to the delimitation of urban sub-areas', *Am. Sociol. Rev.* XIX (1954), 434–40.

FRANKLIN, S. H. 'The age structure of New Zealand's North Island communities', *Economic Geography*, XXXIV (1958), 64–79.

FRANKS, L. K. 'Models for the study of community organization', *Community Dev. Rev.* IX (1958), 1–26.

FRAZER, E. *Home environment and the school* (London, 1959).

FRIED, M. and GLEICHER, P. 'Some sources of residential satisfaction in an urban slum', *J. Am. Inst. of Planners*, XXVII (1961), 305–15.

GANS, H. J. 'Planning and social life: friendship and neighbor relations in sub-communities', *J. Am. Inst. of Planners*, XXVII (1961), 134–40.
'Urbanism and suburbanism as ways of life: a re-evaluation of definitions', in A. Rose (ed.), *Human behaviour and social processes* (London, 1962), pp. 625–48.
The urban villagers (Glencoe, Illinois, 1962).

GARBUTT, G. *A historical and descriptive view of the parishes of Monkwearmouth and Bishopwearmouth and the port and borough of Sunderland* (Sunderland, 1819).

GARRISON, W. L., BERRY, B. J. L., MARBLE, D. F. *et al. Studies of highway development and geographic change* (Seattle, 1959).

GELKHE, C. E. and BIEHL, K. 'Certain effects of grouping upon the size of the correlation coefficient in census tract material', *J. Am. statist. Ass.* XXIX (1934), 169–70.

GENERAL REGISTER OFFICE. *Classification of occupations, 1960* (London, 1960).

GETTYS, W. E. 'Human ecology and social theory', *Social Forces*, XVIII (1940), 469–76. (Also in Theodorson, 1961.)

Bibliography

GITTUS, E. 'The structure of urban areas: a new approach', *Town Planning Review*, XXXV (1964), 5–20.

'An experiment in the definition of urban sub-areas', *Trans. Bartlett Soc.* II (1964–5), 109–35.

'Statistical methods in regional analysis', University of Strathclyde, *Regional Studies Group Bulletin*, no. 3 (1966).

GLASS, R. *The social background of a plan : a study of Middlesbrough* (London, 1948).

GOLDTHORPE, J. H. and LOCKWOOD, D. 'Affluence and the British class structure', *Sociol. Rev.* XI (1963), 133–63.

GOLDTHORPE, J. H., LOCKWOOD, D. *et al.* 'The affluent worker and the thesis of *Embourgeoisement*: some preliminary research findings', *Sociology*, I (1967), 11–31.

GOODMAN, L. A. 'Ecological regressions and the behaviour of individuals', *Am. Sociol. Rev.* XVIII (1953), 663–4.

'Some alternatives to ecological correlation', *Am. J. Sociol.* LXIV (1959), 610–25.

GREEN, N. E. 'Scale analysis of urban structures: Birmingham, Alabama', *Am. Sociol. Rev.* XXI (1956), 8–13.

GREER, S. 'Urbanism reconsidered: a comparative study of local areas in a metropolis', *Am. Sociol. Rev.* XXI (1956), 19–25.

HADDEN, J. K. and BORGATTA, E. F. *American cities : their social characteristics* (Chicago, 1965).

HAGGETT, P. *Locational analysis in human geography* (London, 1965).

HAGOOD, M. J. 'Statistical methods for delineation of regions applied to data on agriculture and population', *Social Forces*, XXI (1943), 288–97.

HAGOOD, M. J., DANILEVSKY, D. and BLUM, C. 'An examination of the use of factor analysis in the problem of subregional delineation', *Rural Sociology*, VI (1941), 216–33.

HARMAN, H. H. *Modern factor analysis* (Chicago, 1960).

HARRIS, C. D. and ULLMAN, E. L. 'The nature of cities', *Ann. Acad. Polit. Soc. Sci.* CCXLII (1945), 7–17. (Also in Mayer and Kohn, 1959.)

HARTSHORNE, R. *The nature of geography* (Chicago, 1939).

Perspective on the nature of geography (Chicago, 1959).

HATT, P. 'The concept of natural area', *Am. Sociol. Rev.* XI (1946), 423–7. (Also in Theodorson, 1961.)

HAWLEY, A. H. *Human ecology : a theory of community structure* (New York, 1950).

HAWLEY, A. H. and DUNCAN, O. D. 'Social area analysis: a critical appraisal', *Land Economics*, XXXIII (1957), 337–45.

HERBERT, D. T. 'An approach to the study of the town as a central place', *Sociological Review*, IX (1961), 273–92.

'Social area analysis: a British study', *Urban Studies*, IV (1967), 41–60.

HIMMELWEIT, H. T. 'Social status and secondary education since the 1944 Act: some data for London', in D. V. Glass (ed.), *Social mobility in Britain* (London, 1954), pp. 141–59.

'Socio-economic background and personality', *Internat. Soc. Sci. Bull.* VII (1955), 29–35.

HOFSTAETTER, P. R. '*Your city* revisited: a factorial study of cultural patterns', *Am. Catholic Sociol. Rev.* XIII (1952), 159–68.

HOLZINGER, K. J. and HARMAN, H. H. *Factor analysis: a synthesis of factorial methods* (Chicago, 1941).

HOTELLING, H. 'Analysis of a complex of statistical variables into principal components', *J. Educ. Psychol.* XXIV (1933), 417–41; 498–520.

HOUSE, J. W. *Recent economic growth in North-East England*. University of Newcastle, Department of Geography Research Series no. 4 (Newcastle, 1964).

HOWE, G. M. *National atlas of disease mortality in the United Kingdom* (London, 1963).

HOYT, H. *The structure and growth of residential neighbourhoods in American cities* (Washington, D.C., 1939). (Pp. 112–22 reprinted in Mayer and Kohn, 1959.) 'Recent distortions of the classical models of urban structure', *Land Economics*, XL (1964), 199–212.

HURD, R. M. *Principles of city land values* (New York, 1903, republished 1924).

HUTCHINSON, W. *The history and antiquities of the County Palatine of Durham* (3 vols. Newcastle, 1785–94).

JACKSON, B. and MARSDEN, D. *Education and the working class* (London, 1962).

JACKSON, W. S. 'Housing and pupil growth and development', *J. of Educ. Sociol.* XXVIII (1955), 370–80.

JAMIESON, J. *Durham at the opening of the twentieth century* (Brighton, 1906).

JAMMER, M. *Concepts of space* (Cambridge, Mass., 1954).

JENNINGS, J. R. G. 'A note on source material for urban geography', *East Midland Geographer*, XX (1963), 212–15.

JOHNSTON, R. J. 'Multi-variate regions: a further approach', *Professional Geographer*, XVII, no. 5 (1965), 9–12. 'The location of high status residential areas', *Geogr. Annlr*, XLVIII B (1966), 23–35.

JONES, E. *A social geography of Belfast* (London, 1960).

JONES, R. 'Segregation in urban residential districts: examples and research problems', in K. Norborg (ed.), *Proc. I.G.U. Symp. Urban Geogr. Lund 1960* (Lund, 1962), pp. 433–46.

KATZ, A. M. and HILL, R. 'Residential propinquity and marital selection: a review of theory, method and fact', *Marriage and Family Living*, XX (1958), 27–35.

KENDALL, M. G. 'The geographical distribution of crop productivity in England', *J.R. statist. Soc.* series A, CII (1939), 21–62. *A course in multivariate analysis* (London, 1957).

KRISTOF, F. S. 'The increased utility of the 1960 housing census for planning', *J. Am. Inst. of Planners*, XXIX (1963), 40–7.

LAMBERT, R. S. *The railway king, 1800–1871: a study of George Hudson and the business morals of his time* (London, 1934).

LAUDER, B. *Towards an understanding of juvenile delinquency* (New York, 1954).

LEWIS, O. 'Further observations on the folk-urban continuum and urbanization with reference to Mexico City', in P. M. Hauser and L. F. Schnore (eds.), *The study of urbanization* (New York, 1965), pp. 491–503.

Bibliography

LOCKWOOD, D. and GOLDTHORPE, J. H. 'The manual worker: affluence, aspiration and assimilation', paper presented to the Annual Meeting of the British Sociological Association, 1962.

LYDALL, H. F. and DAWSON, R. F. F. 'Household income, rent and rates', *Bull. Oxf. Univ. Inst. statist.* XVI (1954), 97–129.

MCCARTY, H. H., HOOK, J. C., and KNOS, D. S. *The measurement of association in industrial geography* (Publications of the Department of Geography, State University of Iowa, no. 1, 1956).

MCELRATH, D. C. 'The social areas of Rome: a comparative analysis', *Am. Sociol. Rev.* XXVII (1962), 376–91.

MCKENZIE, R. D. 'The neighbourhood: a study of local life in the city of Columbus, Ohio', *Am. J. Sociol.* XXVII (1921–2), 145–68, 344–63, 486–508, 588–610, 780–899.

MCQUITTY, L. L. 'Elementary linkage analysis for isolating orthogonal and oblique types and typal relevancies', *Educ. Psychol. Meas.* XVII (1957), 207–29.

MARTIN, F. M. 'Parents' preferences in secondary education', in D. V. Glass (ed.), *Social mobility in Britain* (London, 1954), pp. 160–74.

MAYFIELD, R. C. 'Conformations of service and retail activities', in K. Norborg (ed.), *Proc. I.G.U. Symp. Urban Geogr. Lund 1960* (Lund, 1962), pp. 77–89.

MAYS, J. B. *Education and the urban child* (Liverpool, 1962).

MEIER, R. L. *A communication theory of urban growth* (Harvard, 1962).

MENZEL, H. 'Comment on Robinson's "ecological correlation and the behaviour of individuals"', *Am. Sociol. Rev.* XV (1950), 674.

MINISTRY OF HOUSING AND LOCAL GOVERNMENT. *Report of the committee on housing in Greater London* (Cmd. 2605, London, 1965). (Milner Holland Report.)

The housing programme, 1965–70. (Cmnd. 2838, London, 1965.)

MINISTRY OF LABOUR. *Gazette*, monthly issues, 1954–65.

MITCHELL, G. D., LUPTON, T. *et al. Neighbourhood and Community* (Liverpool, 1954).

MITCHELL, W. C. *A History of Sunderland* (Sunderland, 1919).

MOGEY, J. M. *Family and Neighbourhood: two studies in Oxford* (London, 1956).

MORGAN, W. B. and MOSS, R. P. 'Geography and Ecology: the concept of the community and its relationship to environment', *Ann. Ass. Am. Geogr.* LV (1965), 339–50.

MORRILL, R. and PITTS, F. R. 'Marriage, migration and the mean information field: a study in uniqueness and generality', *Ann. Ass. Am. Geogr.* LVII (1967), 401–22.

MORRIS, R. N. and MOGEY, J. M. *The sociology of housing: studies at Berinsfield* (London, 1965).

MORRIS, T. *The criminal area: a study in social ecology* (London, 1957).

MOSER, C. A. *Survey methods in social investigation* (London, 1958).

MOSER, C. A. and SCOTT, W. *British towns: a statistical study of their social and economic differences* (London, 1961).

MYERS, J. K. 'Note on the homogeneity of census tracts', *Social Forces*, XXXII (1954), 364–6.

NATIONAL ACADEMY OF SCIENCES–NATIONAL RESEARCH COUNCIL. *The science of geography*, report of the committee on geography (Washington, D.C., 1965).

NEF, J. U. *The rise of the British coal industry* (London, 1932).

OXFORD CENSUS TRACT COMMITTEE. *Census 1951—Oxford area : selected population and housing characteristics by census tracts* (Oxford, 1957).

PARK, R. E. 'The City: suggestions for the investigation of human behaviour in the urban environment', *Am. J. Sociol.* XX (1916), 577–612.

The collected papers of Robert Ezra Park. I, *Race and culture* (Glencoe, Illinois, 1950); II, *Human communities* (Glencoe, 1952); III, *Society* (Glencoe, 1955).

PARK, R. E., BURGESS, E. W. and MCKENZIE, R. D. *The City* (Chicago, 1925, reprinted 1967).

PLANT, J. S. 'The personality and an urban area', in P. K. Hatt and A. J. Reiss (eds.), *Cities and Society* (2nd ed. Glencoe, 1957), pp. 647–65.

PLATT, R. S. 'Field approach to regions', *Ann. Ass. Am. Geogr.* XXV (1935), 153–74.

POTTS, T. *Sunderland : a history of the town, port, trade and commerce* (Sunderland, 1892).

PRICE, D. O. 'Factor analysis in the study of metropolitan centres', *Social Forces*, XX (1942), 449–55.

QUINN, J. A. 'The Burgess zonal hypothesis and its critics', *Am. Sociol. Rev.* V (1940), 210–18.

Human Ecology (New York, 1950).

RAMSØY, N. R. 'Assortive mating and the structure of cities', *Am. Sociol. Rev.* XXXI (1966), 773–86.

RATCLIFFE, R. U. 'The dynamics of efficiency in the locational distribution of urban activities', in R. M. Fisher (ed.), *The metropolis in modern life* (New York, 1955), pp. 125–48. (Also in Mayer and Kohn, 1959.)

REDFIELD, R. *The folk culture of Yucatan* (Chicago, 1941).

REID, M. G. *Housing and Income* (Chicago, 1962).

REISSMAN, L. *The urban process : cities in industrial society* (New York, 1964).

ROBINSON, A. H. 'The necessity of weighting values in correlation analysis of areal data', *Ann. Ass. Am. Geogr.* XLVI (1956), 233–6.

ROBINSON, W. S. 'Ecological correlation and the behaviour of individuals', *Am. Sociol. Rev.* XV (1950), 351–7.

ROSSI, P. H. 'Comment', *Am. J. Sociol.* LXV (1959), 146–9.

RUNCIMAN, W. G. *Relative deprivation and social justice : a study of attitudes to social inequality in twentieth-century England* (London, 1966).

Ryde on Rating, see under 'Williams'.

SCHMID, C. F. 'The theory and practice of planning census tracts', *Sociol. Soc. Res.* XXII (1938), 228–38.

'Generalizations concerning the ecology of the American city', *Am. Sociol. Rev.* XV (1950), 264–81.

Bibliography

SCHMID, C. F., MCCANNELL, E. H. and VAN ARSDOL, M. D., Jr. 'The ecology of the American City: further comparison and validation of generalizations', *Am. Sociol. Rev.* XXIII (1958), 392–401. (Also in Theodorson, 1961.)

SCHNORE, L. F. 'Social morphology and human ecology', *Am. J. Sociol.* LXIII (1958), 620–34.

'The myth of human ecology', *Sociol. Inquiry*, XXXI (1961), 128–39.

'Geography and human ecology', *Econ. Geogr.* XXXVII (1961), 207–17.

'A planner's guide to the 1960 census of population', *J. Am. Inst. of Planners* XXIX (1963), 29–39.

The urban scene : human ecology and demography (New York, 1965).

SCHORR, A. L. *Slums and social insecurity* (London, 1964).

SHAW, C. R. *Delinquency areas* (Chicago, 1929).

SHEVKY, E. and BELL, W. *Social area analysis : theory, illustrative application and computational procedure* (Stanford, 1955).

SHEVKY, E. and WILLIAMS, M. *The social areas of Los Angeles : analysis and typology* (Berkeley, 1949).

SJOBERG, G. *The preindustrial city : past and present* (New York, 1960).

'On the spatial structure of cities in the two Americas', in P.M. Hauser and L. F. Schnore (eds.), *The study of urbanization* (New York, 1965), pp. 347–98.

'Theory and research in urban sociology', in *ibid.* pp. 157–89.

SMAILES, A. E. 'Early industrial settlement in North-East England', *Advmt Sci.* XVI (1950), 325–31.

The geography of towns (London, 1953).

'Greater London: the structure of a metropolis', *Geogr. Z.* LII (1964), 163–89.

SMITH, J. 'A method for the classification of areas on the basis of demographically homogeneous populations', *Am. Sociol. Rev.* XIX (1954), 201–7.

SMITH, J. and MADDOX, G. L. 'The spatial location and the use of selected facilities in a middle-sized city', *Social Forces*, XXXVIII (1959), 119–24.

SMITH, J. W. and HOLDEN, T. S. *Where ships are born : a history of shipbuilding on the River Wear* (Sunderland, 1953).

SMITH, R. H. T. 'The functions of Australian towns', *Tijdschr. econ. Sociale Geogr.* LVI (1965), 81–92.

SOCIAL SURVEY. *The proportion of jurors as an index of the economic status of a district.* (P. G. Gray et al.) (London, 1951).

SPROUT, H. and SPROUT, M. *The ecological perspective on human affairs : with special reference to international politics* (Princeton, 1965).

STACEY, M. *Tradition and change : a study of Banbury* (London, 1960).

STEWART, J. Q. and WARNTZ, W. 'Physics of population distribution', *J. Reg. Sci.* I (1958), 99–123.

STODDART, D. R. 'Geography and the ecological approach: the ecosystem as a geographic principle and method', *Geography*, L (1965), 242–51.

STONE, J. R. N. 'The interdependence of blocks of transactions', *J. R. statist. Soc. Suppl.* IX (1947), 1–32.

STOUFFER, S. A., GUTTMAN, L., SUCHMAN, E. A. and LAZARSFELD, P. *Studies in social psychology in World War II*, IV, *Measurement and prediction* (Princeton, 1950).

SWEETSER, F. L. 'Factorial ecology, Helsinki, 1960', *Demography*, II (1965), 372–85. 'Factor structure as ecological structure in Helsinki and Boston', *Acta Sociologica*, VIII (1965), 205–25.

TAYLOR, W. P. 'What is ecology and what good is it?', *Ecology*, XVII (1936), 333–46.

THOMAS, E. N. and ANDERSON, D. L. 'Additional comments on weighting values in correlation analysis of areal data', *Ann. Ass. Am. Geogr.* LV (1965), 492–505.

THOMPSON, J. H., SUFRIN, S. C., GOULD, P. R. and BUCK, M. A. 'Toward a geography of economic health', *Ann. Ass. Am. Geogr.* LII (1962), 1–20.

THURSTONE, L. L. *Multiple-factor analysis : a development and expansion of the 'vectors of mind'* (Chicago, 1947).

TIMMS, D. W. G. 'Quantitative techniques in urban social geography', in R. J. Chorley and P. Haggett (eds.), *Frontiers in geographical teaching* (London, 1965), pp. 257–61.

TOMLINSON, W. W. *The North-East Railway : its rise and development.* (Newcastle, 1914).

TREWARTHA, G. T. 'The case for population geography', *Ann. Ass. Am. Geogr.* XLIII (1953), 71–97.

TRYON, R. C. *Identification of social areas by cluster analysis* (Berkeley, 1955).

U.S. BUREAU OF THE CENSUS. 'The 1950 censuses: how they were taken', *Procedural Studies of the 1950 Censuses*, no. 2 (Washington, 1955). *Census Tract Manual* (4th ed. Washington, 1958).

VAN ARSDOL, M. D. Jr., CAMILLERI, S. F. and SCHMID, C. F. 'The generality of urban social area indexes', *Am. Sociol. Rev.* XXIII (1958), 277–84. (Also in Theodorson, 1961.) 'An application of the Shevky social area indexes to a model of urban society', *Social Forces*, XXXVII (1958), 26–32.

Victoria History of the County of Durham (W. Page, ed.). 3 vols. (London, 1905–28).

VON BERTALANFFY, L. 'An outline of general system theory', *Br. J. Phil. Sci.* I (1950), 134–65.

VON STRUVE, A. W. 'Geography in the census bureau', *Economic Geography*, XVI (1940), 275–80.

WARNTZ, W. *Geography now and then : some notes on the history of academic geography in the United States* (New York, 1964).

WATERS, R. A. *Hendon : past and present* (Sunderland, 1900).

WEAVER, J. C. 'Crop combination regions in the Middle West', *Geogr. Rev.* XLIV (1954), 175–200.

WHYTE, W. F. *Street corner society : the social structure of an Italian slum* (Chicago, 1943, 2nd ed. 1955).

WHYTE, W. H. Jr. *The organization man* (London, 1957).

WILLIAMS, H. B. *et al.* (eds.). *Ryde on rating : the law and the practice.* (11th ed. London, 1963).

Bibliography

WILLMOTT, P. *The evolution of a community : a study of Dagenham after forty years* (London, 1963).

WILLMOTT, P. and YOUNG, M. *Family and class in a London suburb* (London, 1960).

WILNER, D. M., WALKLEY, R. P. *et al. The housing environment and family life : a longitudinal study of the effects of housing on morbidity and mental health* (Baltimore, 1962).

WILSON, A. B. 'Residential segregation of social classes and aspirations of high school boys', *Am. Sociol. Rev.* XXIV (1959), 836–45.

WIRTH, L. 'Urbanism as a way of life', *Am. J. Sociol.* XLIV (1938), 1–24. (Also in Hatt and Reiss, 1957.)

WRIGLEY, E. A. 'Geography and Population', in R. J. Chorley and P. Haggett (eds.), *Frontiers in geographical teaching* (London, 1965), pp. 62–80.

YOUNG, M. and WILLMOTT, P. *Family and kinship in East London* (London, 1957).

ZELINSKY, W. *A bibliographic guide to population geography.* University of Chicago, Department of Geography Research Paper no. 80 (Chicago, 1962).

ZOBLER, L. 'Decision making in regional construction', *Ann. Ass. Am. Geogr.* XLVIII (1958), 140–8.

ZORBAUGH, H. W. 'The natural areas of the city', *Publ. Pap. Am. Sociol. Soc.* XX (1926), 188–97. (Also in Theodorson, 1961.)

INDEX

Index

Milton Keynes UK
Ingram Content Group UK Ltd.
UKHW041523181024
449640UK00009B/167